Perturbation Methods

Cambridge texts in applied mathematics

Maximum and minimum principles: a unified approach with applications
M.J. SEWELL

Introduction to numerical linear algebra and optimization
P.G. CIARLET

Solitons: an introduction
P.G. DRAZIN AND R.S. JOHNSON

Integral equations: from spectral theory to applications
D. PORTER AND DAVID S.G. STIRLING

Perturbation methods
E.J. HINCH

Perturbation Methods

E.J. HINCH

Lecturer in Mathematics, University of Cambridge

Published by the Press Syndicate of the University of Cambridge
The Pitt Building, Trumpington Street, Cambridge CB2 1RP
40 West 20th Street, New York, NY 10011-4211, USA
10 Stamford Road, Oakleigh, Melbourne 3166, Australia

© Cambridge University Press 1991

First published 1991
Reprinted 1992, 1994, 1995

Library of Congress Cataloging-in-Publication Data is available.

A catalogue record for this book is available from the British Library.

ISBN 0-521-37310-7 hardback
ISBN 0-521-37897-4 paperback

Transferred to digital printing 2002

Contents

Preface

Making precise approximations to solve equations is an occupation of applied mathematicians which distinguishes them from pure mathematicians, physicists and engineers. A precise approximation is not a contradiction in terms but rather an approximation with an error which is understood and controllable; in particular the error could be made smaller by some rational procedure. There are two methods for obtaining precise approximations to the solutions of an equation, numerical methods and analytic methods, and this book is about the latter. The analytic approximations are obtained when some parameter of the problem is small, and hence the name *perturbation methods*. The perturbation and numerical methods are not however in competition but rather complement one another as the following example illustrates.

The van der Pol oscillator is governed by the equation

$$\ddot{x} + k\dot{x}(x^2 - 1) + x = 0$$

In time the solution tends to an oscillation with a particular amplitude which does not depend on the initial conditions. The period of this limit oscillation is of interest and is plotted in figure 1 as a function of the strength of the nonlinear friction, k. The circles give the numerical results obtained by a Runge–Kutta method. The dashed curves give the first and second order perturbation approximations

$$\text{Period} = \begin{cases} 2\pi \left(1 + \frac{1}{16}k^2 + O(k^4)\right) & \text{as } k \to 0 \\ k(3 - 2\ln 2) + 7.0143k^{-1/3} + O(k^{-1}\ln k) & \text{as } k \to \infty \end{cases}$$

At intermediate values of the parameter k, from 2 to 6, the numerical method is most useful. At extreme values however the numerical method loses its accuracy rapidly, for example by $k = 10$ the time-step must be reduced to 0.01 in order to obtain 5 figure accuracy. The analytic approximations take over in the extreme conditions. Further they give an explicit dependence on the parameter k rather than the isolated results

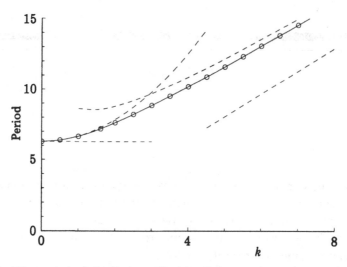

Fig. 1   The period of the limit oscillation of the van der Pol oscillator as a function of the strength of the nonlinear friction k.

at particular values from the numerical method. But the most important feature of the figure is the satisfying agreement between the numerical approximation and the two independent perturbation approximations – such checks are essential in research.

Obtaining good numerical values for the solution is not the only quest of a perturbation approximation. One can hope that the analysis will reveal some physical insights through the simplified physics of the limiting problem. In this book I will however suppress the physics in the problems discussed.

Finding perturbation approximations is an art rather than a science. In research it is useful to be responsive to suggestions from the physics. There is certainly no routine method appropriate to all problems, or even classes of problems. Instead one needs a determination to exploit the smallness of the parameter. This book attempts to present many of the weapons which have been found useful, but they should not be viewed as exhaustive.

While this book is mathematical, no attempt has been made to make the arguments fully rigorous. In general I have tried to explain why the results are correct. Often these reasons can be turned into strict theorems, albeit with some difficulty in the case of singular problems. My own opinion is that such superficial rigour rarely adds to the understanding of the problem, and that of greater use is a numerical statement about the range of applicability achieving some specified accuracy.

This book is based on a course of lectures which I gave for a number of years to first year graduate students in the University of Cambridge. In its turn it was based on my own education from a course of lectures by L. E. Fraenkel and from the book on the subject by M. Van Dyke. These two inspiring teachers asked many interesting questions which I have attempted to answer in this book; questions such as why are some results convergent whilst others only asymptotic, why is matching possible, what selection criterion should be used with strained co-ordinates, and what characterises problems to be tackled by multiple scales.

While no previous knowledge of perturbation methods is assumed, some previous experience is probable. The students who attended my lecture course would have seen several examples (small friction on projectiles, perturbed energy levels in quantum mechanics, adiabatic invariants in Hamiltonian systems, Watson's lemma, and viscous boundary layers in fluid mechanics) usually presented in an informal way relying heavily on physical insight. They would not however have seen a formal and organised approach to a perturbation problem.

The eventual goal of this book is to present the method of matched asymptotic expansions and the method of multiple scales, progressing to an advanced level in considering the more difficult issues such as the occurrence of logarithms and the occurrence of more than two scales. Tackling differential equations with such singular perturbation problems is certainly not easy. Fortunately many of the essential concepts can be presented in the simpler context of algebraic equations and later with integrals. Thus issues such as iterations and expansions, singular problems and rescaling, non-integral powers and logarithms will be presented well before the difficult singular differential equations are encountered. Finally I should observe that most of the chapters follow the basic method with an advanced application whose understanding is not essential to the following chapters – thus §§ 1.6, 3.5, 5.3, 5.4, 5.5, 5.6, 6.3, 7.3, 7.4 and 7.6 should be viewed as optional.

E.J. Hinch
Cambridge, 1990

Algebraic equations

Many of the techniques of perturbation analysis can be introduced in the simple setting of algebraic equations. By starting with some particularly easy algebraic equations, three quadratics, we can benefit from the luxury of the existence of exact answers, taking useful hints from them to overcome difficulties.

1.1   Iteration and expansion

We start with the equation for x which contains the parameter ϵ,

$$x^2 + \epsilon x - 1 = 0$$

This has exact solutions

$$x = -\tfrac{1}{2}\epsilon \pm \sqrt{1 + \tfrac{1}{4}\epsilon^2}$$

which can be expanded for small ϵ as

$$x = \begin{cases} +1 - \tfrac{1}{2}\epsilon + \tfrac{1}{8}\epsilon^2 - \tfrac{1}{128}\epsilon^4 + O(\epsilon^6) \\ -1 - \tfrac{1}{2}\epsilon - \tfrac{1}{8}\epsilon^2 + \tfrac{1}{128}\epsilon^4 + O(\epsilon^6) \end{cases}$$

These binomial expansions converge if $|\epsilon| < 2$.

More important than converging, the truncated series give a good approximation if ϵ is small. The first few terms give a result within 3% of the exact result if

$$|\epsilon| \quad < \quad 0.05 \qquad 0.5 \qquad 1.2 \qquad 1.6$$

$$\vdots \qquad\quad \vdots \qquad\quad \vdots \qquad\quad \vdots$$

$$x \quad = \quad 1 \quad - \quad \tfrac{1}{2}\epsilon \quad + \quad \tfrac{1}{8}\epsilon^2 \quad - \quad \tfrac{1}{128}\epsilon^4 \quad + \quad O(\epsilon^6)$$

The last 1.6 being not too far from the convergence boundary. Alternatively for the fixed value of $\epsilon = 0.1$ the first few terms give

$$x \quad \sim \quad 1.0$$
$$0.95$$
$$0.95125$$
$$0.95124921\ldots$$
$$\ldots$$
$$\text{exact} \quad = \quad 0.95124922\ldots$$

Often the numerical summation of these short expansions involves less computer time than the evaluation of the exact answer with its costly surds.

We started by finding the exact solution of the quadratic equation and then we expanded the exact solution. In most problems, however, it is not possible to find the exact solution. We must therefore develop techniques which first make the approximations and then, only afterwards, involve a calculation. There are two distinct methods of first approximating and then calculating, the iterative method and the expansion method. Each method has its own advantages and disadvantages.

Iterative method

We start with the iterative method, because it is a method which is often overlooked although it has much to offer.

The first step of the iterative method is to find a *rearrangement* of the original equation which will become the basis of an iterative process. This first step involves a certain amount of inspiration which must therefore count as a major drawback of the method. A suitable rearrangement of our present quadratic is

$$x \; = \; \pm\sqrt{1 - \epsilon x}$$

Any solution of the original equation is a solution of this rearrangement and vice versa.

Working with just the positive root, we thus adopt the iterative process

$$x_{n+1} \; = \; \sqrt{1 - \epsilon x_n}$$

The iterative process needs a starting point, the value of the root when $\epsilon = 0$, $x_0 = 1$.

Making the first iteration, we find

$$x_1 = \sqrt{1 - \epsilon}$$

which can be expanded in a binomial series

$$x_1 = 1 - \tfrac{1}{2}\epsilon - \tfrac{1}{8}\epsilon^2 - \tfrac{1}{16}\epsilon^3 + \cdots$$

Looking at the exact answer, we see that the ϵ^2 and higher terms are erroneous. We therefore truncate the series for x_1 after the second term:

$$x_1 = 1 - \tfrac{1}{2}\epsilon + \cdots$$

Proceeding to the next iteration, we find

$$x_2 = \sqrt{1 - \epsilon(1 - \tfrac{1}{2}\epsilon)}$$

which can be expanded, this time retaining only terms up to ϵ^2:

$$
\begin{aligned}
x_2 &= 1 - \tfrac{1}{2}\epsilon(1 - \tfrac{1}{2}\epsilon) - \tfrac{1}{8}\epsilon^2(1 + \cdots)^2 + \cdots \\
&= 1 - \tfrac{1}{2}\epsilon + \tfrac{1}{8}\epsilon^2 + \cdots
\end{aligned}
$$

We note that the ϵ^2 term is now correct after two iterations. Iterating again, we find

$$
\begin{aligned}
x_3 &= \sqrt{1 - \epsilon(1 - \tfrac{1}{2}\epsilon + \tfrac{1}{8}\epsilon^2)} \\
&= 1 - \tfrac{1}{2}\epsilon(1 - \tfrac{1}{2}\epsilon + \tfrac{1}{8}\epsilon^2) - \tfrac{1}{8}\epsilon^2(1 - \tfrac{1}{2}\epsilon + \cdots)^2 - \tfrac{1}{16}\epsilon^3(1 - \cdots)^3 + \cdots \\
&= 1 - \tfrac{1}{2}\epsilon + \tfrac{1}{8}\epsilon^2 + 0\epsilon^3 + \cdots
\end{aligned}
$$

It is clear that progressively more work is required to obtain the higher order terms by the iterative method. The method also has the undesirable feature that in the early iterations it gives erroneous values to the higher terms. One can only check that a term is correct by making one more iteration, which of course is usually convincing but no rigorous proof (but see §1.5).

Expansion method

The first step of the expansion method is to set $\epsilon = 0$ and find the unperturbed roots $x = \pm 1$. Then one poses an expansion about one of these roots, say $x = +1$, expanding in powers of ϵ, i.e.

$$x(\epsilon) = 1 + \epsilon x_1 + \epsilon^2 x_2 + \epsilon^3 x_3 + \cdots$$

This expansion is formally substituted into the governing quadratic equation.

$$1 + \epsilon 2x_1 + \epsilon^2(x_1^2 + 2x_2) + \epsilon^3(2x_1x_2 + 2x_3) + \cdots$$
$$+ \epsilon + \epsilon^2 x_1 + \epsilon^3 x_2 + \cdots$$
$$-1$$
$$= 0$$

Here one ignores potential difficulties in making the substitution such as the limitations in multiplying series term by term. The coefficients of the powers of ϵ on the two sides of the equation are now compared.

At ϵ^0: $1 - 1 = 0$

This level is satisfied automatically because we started the expansion from the correct value $x = 1$ at $\epsilon = 0$.

At ϵ^1: $2x_1 + 1 = 0$ i.e. $x_1 = -\frac{1}{2}$

At ϵ^2: $x_1^2 + 2x_2 + x_1 = 0$ i.e. $x_2 = \frac{1}{8}$

Here the previously determined value of x_1 has been used.

At ϵ^3: $2x_1x_2 + 2x_3 + x_2 = 0$ i.e. $x_3 = 0$

again using the previously determined values of x_1 and x_2.

The expansion method is much easier than the iterative method when working to higher orders. To use the expansion method, however, it is necessary to assume that the result can be expanded in powers of ϵ and that the formal substitution and associated manipulations are permitted.

Exercise 1.1. Find four terms in the expansion of the root near $x = -1$, using both the iterative and expansion methods.

1.2   Singular perturbations and rescaling

In this section we study the quadratic

$$\epsilon x^2 + x - 1 = 0$$

If $\epsilon = 0$ there is just one root, at $x = 1$, whereas when $\epsilon \neq 0$ there are two roots. This is an example of a *singular perturbation* problem, in which the limit point $\epsilon = 0$ differs in an important way from the approach to the limit $\epsilon \to 0$. Interesting problems are often singular. Problems which are not singular are said to be *regular*.

To resolve the paradox of the behaviour of the second root we take the exact solutions to the quadratic and expand them for small ϵ (convergent if $|\epsilon| < \frac{1}{4}$). The two roots are

$$x \;=\; \begin{cases} 1 - \epsilon + 2\epsilon^2 - 5\epsilon^3 + \cdots \\ -1/\epsilon - 1 + \epsilon - 2\epsilon^2 + 5\epsilon^3 + \cdots \end{cases}$$

Thus the singular second root evaporates off to $x = \infty$ in the limit $\epsilon = 0$.

Iterative method

To set up an iterative process for the singular root we argue as follows. In order to retain the second solution of the governing quadratic, it is necessary to keep the ϵx^2 term as a main term rather than as a small correction. Thus x must be large. Hence at leading order the -1 term in the equation will be negligible when compared with the x term, i.e.

$$\epsilon x^2 + x \;\approx\; 0 \qquad \text{with solution} \qquad x \sim -1/\epsilon$$

Hence we are led to the rearrangement of the quadratic

$$x \;=\; -\frac{1}{\epsilon} + \frac{1}{\epsilon x}$$

and the iterative process

$$x_{n+1} \;=\; -\frac{1}{\epsilon} + \frac{1}{\epsilon x_n}$$

with a starting point $x_0 = -1/\epsilon$.
  Iterating once we find

$$x_1 \;=\; -\epsilon^{-1} - 1$$

and iterating again

$$x_2 \;=\; -\epsilon^{-1} - \frac{1}{1+\epsilon}$$
$$=\; -\epsilon^{-1} - 1 + \epsilon + \cdots$$

A further iteration is needed to obtain the ϵ^2 term correctly.

Expansion method

The expansion method can be applied to the singular root by posing a power series in ϵ which starts with an ϵ^{-1} term instead of the usual ϵ^0. The way in which one determines the correct starting point is left until later in this section. Thus substituting

$$x(\epsilon) \;=\; \epsilon^{-1} x_{-1} + x_0 + \epsilon x_1 + \cdots$$

into the governing quadratic yields

$$\begin{aligned}
\epsilon^{-1}x_{-1}^2 &+ 2x_{-1}x_0 &+ \epsilon(2x_{-1}x_1 + x_0^2) + \cdots \\
\epsilon^{-1}x_{-1} &+ x_0 &+ \epsilon x_1 + \cdots \\
&- 1 & \\
&= 0 &
\end{aligned}$$

Comparing coefficients of ϵ^n, we have

at ϵ^{-1}: $x_{-1}^2 + x_{-1} = 0$,   i.e. $x_{-1} = -1$ or 0

The root $x_{-1} = 0$ leads to the regular root, so we ignore it.

At ϵ^0: $2x_{-1}x_0 + x_0 - 1 = 0$,   i.e. $x_0 = -1$

At ϵ^1: $2x_{-1}x_1 + x_0^2 + x_1 = 0$,   i.e. $x_1 = 1$

where at each stage the values of previously determined x_n have been used.

Rescaling in the expansion method

Instead of starting the expansion with the unusual ϵ^{-1} term, a very useful idea for singular problems is to rescale the variables before making the expansion. Thus introducing the rescaling

$$x = X/\epsilon$$

into the originally singular equation for x produces an equation for X,

$$X^2 + X - \epsilon = 0$$

which is regular. Thus the problem of finding the correct starting point for the expansion can be viewed as a problem of finding a suitable rescaling to regularise the singular problem.

There is a simple procedure to find all useful rescalings. First one poses a general rescaling with a scaling factor $\delta(\epsilon)$,

$$x = \delta X$$

in which one insists that X is strictly of order unity as $\epsilon \to 0$. Unfortunately the standard notation $X = O(1)$ does not describe this limitation on X, because $O(1)$ permits X to be vanishingly small as $\epsilon \to 0$. Thus we are forced to adopt the less familiar notation $X = \text{ord}(1)$ to stand for X *is strictly of order unity as* $\epsilon \to 0$.

Substituting the general rescaling into the governing quadratic equation gives

$$\epsilon\delta^2 X^2 + \delta X - 1 = 0$$

We now consider the dominant balance of this equation for δ of different magnitudes, starting the search for sensible rescalings with δ very small and progressing to δ very large.

- $\delta \ll 1$. If δ is very small, then the left hand side of the equation is

$$\epsilon \delta^2 X^2 + \delta X - 1 \quad = \quad \text{small} + \text{small} - 1$$

This cannot balance the zero on the right hand side, and so a small δ is an unacceptable rescaling. As δ is increased, it is the X term which first breaks the domination of the 1 term and this occurs when $\delta = 1$. Hence the range of the unacceptable rescalings when δ is too small is $\delta \ll 1$, as declared above.

- $\delta = 1$. The left hand side of the equation is now

$$\epsilon \delta^2 X^2 + \delta X - 1 \quad = \quad \text{small} + X - 1$$

This can balance the zero on the right hand side to produce the regular root $X = 1 + \text{small}$.

- $1 \ll \delta \ll \epsilon^{-1}$. If δ is a little larger than unity, then the X term dominates the left hand side of the equation

$$(\epsilon \delta^2 X^2 + \delta X - 1)/\delta \quad = \quad \text{small} + X + \text{small}$$

This can balance the zero divided by δ on the right hand side, but only if $X = 0 + \text{small}$ which violates the restriction that X is strictly of order unity and not smaller. This rescaling is therefore unacceptable. As δ increases well beyond unity, it is the X^2 term which breaks the domination of the X term when $\delta = \epsilon^{-1}$. Hence this range of unacceptable rescalings is $1 \ll \delta \ll \epsilon^{-1}$, as declared above.

- $\delta = \epsilon^{-1}$. The left hand side of the equation divided by $\epsilon \delta^2$ is

$$(\epsilon \delta^2 X^2 + \delta X - 1)/\epsilon \delta^2 \quad = \quad X^2 + X + \text{small}$$

This can balance the zero divided by $\epsilon \delta^2$ on the right hand side with either $X = -1 + \text{small}$ which yields the singular root, or $X = 0 + \text{small}$ which is not permitted because it violates our restriction $X = \text{ord}(1)$.

- $\epsilon^{-1} \ll \delta$. Finally when δ is very large, the left hand side of the quadratic divided by $\epsilon \delta^2$ is

$$(\epsilon \delta^2 X^2 + \delta X - 1)/\epsilon \delta^2 \quad = \quad X^2 + \text{small} + \text{small}$$

This can only balance the right hand side if $X = 0 + \text{small}$, which violates $X = \text{ord}(1)$. Thus $\epsilon^{-1} \ll \delta$ is a range of unacceptable rescalings.

The systematic search of all possible rescalings has thus yielded $\delta = 1$ for the regular root and $\delta = \epsilon^{-1}$ for the singular root as the only possible rescalings with $X = \text{ord}(1)$.

Exercise 1.2.   Find the rescalings for the roots of

$$\epsilon^2 x^3 + x^2 + 2x + \epsilon = 0$$

and thence find two terms in the approximation for each root.

1.3   Non-integral powers

In this section we study the quadratic

$$(1 - \epsilon)x^2 - 2x + 1 = 0$$

This innocent looking equation gives an unexpected surprise.

We start this time with the expansion method. Setting $\epsilon = 0$ we
have the unperturbed solution $x = 1$, a double root. One learns from
experience that a multiple root is a sign of imminent danger. Proceeding
however in the usual manner, we pose the expansion in powers of ϵ

$$x(\epsilon) = 1 + \epsilon x_1 + \epsilon^2 x_2 + \cdots$$

Substituting into the governing equation

$$
\begin{aligned}
1 \quad &+ \quad \epsilon 2x_1 \quad + \quad \epsilon^2(2x_2 + x_1^2) \quad + \cdots \\
&- \quad \epsilon \quad\quad - \quad \epsilon^2 2x_1 \quad\quad\quad\ + \cdots \\
-2 \quad &- \quad \epsilon 2x_1 \quad - \quad \epsilon^2 2x_2 \quad\quad\quad + \cdots \\
+1 \\
= \quad 0 &
\end{aligned}
$$

and comparing coefficients of ϵ^n we find

At ϵ^0: $1 - 2 + 1 = 0$

which is automatically satisfied because we started correctly perturbing
about $x = 1$.

At ϵ^1: $2x_1 - 1 - 2x_1 = 0$

This cannot be satisfied with any value of x_1, except in some sense with
$x_1 = \infty$.

To find the cause of the difficulty we look at the exact solution of the
quadratic

$$x = \frac{1 \pm \epsilon^{\frac{1}{2}}}{1 - \epsilon}$$

Taking just the positive root and expanding for small ϵ, we find

$$x = 1 + \epsilon^{\frac{1}{2}} + \epsilon + \epsilon^{\frac{3}{2}} + \cdots$$

Thus we see that we should have expanded in powers of $\epsilon^{\frac{1}{2}}$ rather than powers of ϵ. This is what the infinite value that we found above for x_1 was hinting: in a certain sense $\epsilon^{\frac{1}{2}} = \epsilon \times \infty$. In retrospect we could also have foreseen that an $O(\epsilon^{\frac{1}{2}})$ change in the variable would be required to produce an $O(\epsilon)$ change in a function at its minimum. Returning to the quadratic, we now pose an expansion in powers of the unexpected non-integral powers

$$x(\epsilon) = 1 + \epsilon^{\frac{1}{2}}x_{\frac{1}{2}} + \epsilon x_1 + \epsilon^{\frac{3}{2}}x_{\frac{3}{2}} + \cdots$$

Substituting this into the governing equation

$$
\begin{aligned}
1 \quad &+ \quad \epsilon^{\frac{1}{2}}2x_{\frac{1}{2}} \quad + \quad \epsilon(2x_1 + x_{\frac{1}{2}}^2) \quad + \quad \epsilon^{\frac{3}{2}}(2x_{\frac{3}{2}} + 2x_{\frac{1}{2}}x_1) \quad + \cdots \\
&\qquad\qquad\qquad\quad - \quad \epsilon \qquad\qquad\qquad - \quad \epsilon^{\frac{3}{2}}2x_{\frac{1}{2}} \qquad\qquad + \cdots \\
-2 \quad &- \quad \epsilon^{\frac{1}{2}}2x_{\frac{1}{2}} \quad - \quad \epsilon 2x_1 \qquad\qquad - \quad \epsilon^{\frac{3}{2}}2x_{\frac{3}{2}} \qquad\qquad + \cdots \\
+1 \quad & \\
= 0 &
\end{aligned}
$$

Comparing coefficients of $\epsilon^{\frac{n}{2}}$ we find that as usual

at ϵ^0: $1 - 2 + 1 = 0$

is automatically satisfied and that

at $\epsilon^{\frac{1}{2}}$: $2x_{\frac{1}{2}} - 2x_{\frac{1}{2}} = 0$

This is satisfied by all values of $x_{\frac{1}{2}}$. It is a little disturbing that $x_{\frac{1}{2}}$ has not been determined at the $\epsilon^{\frac{1}{2}}$ level, but we proceed to the next level.

At ϵ^1: $2x_1 + x_{\frac{1}{2}}^2 - 1 - 2x_1 = 0$

So $x_{\frac{1}{2}} = \pm 1$ and x_1 is not determined at this level. Continuing to the next level

at $\epsilon^{\frac{3}{2}}$: $2x_{\frac{3}{2}} + 2x_{\frac{1}{2}}x_1 - 2x_{\frac{1}{2}} - 2x_{\frac{3}{2}} = 0$

So $x_1 = 1$ for both roots of $x_{\frac{1}{2}}$, while $x_{\frac{3}{2}}$ is not determined.

The delay in determining $x_{\frac{n}{2}}$ at the $\epsilon^{\frac{n+1}{2}}$ level rather than at the $\epsilon^{\frac{n}{2}}$ level means that a little extra work is required. There is also the slight worry that at the following level it will not be possible to satisfy the equation and the whole solution will therefore collapse, as happened at the ϵ level in the erroneous expansion in powers of ϵ.

Finding the expansion sequence

Having rescued the expansion method by looking at the exact answer, there remains the problem of how one determines the expansion sequence when the exact answer is not available.

First one poses a general expansion

$$x(\epsilon) \;=\; 1 + \delta_1(\epsilon)x_1 + \delta_2(\epsilon)x_2 + \cdots$$

where one requires

$$1 \gg \delta_1(\epsilon) \gg \delta_2(\epsilon) \gg \ldots \quad \text{and} \quad x_1, x_2, \ldots = \mathrm{ord}(1) \text{ as } \epsilon \to 0$$

Substituting into the governing quadratic yields

$$1 + 2\delta_1 x_1 + \delta_1^2 x_1^2 + 2\delta_2 x_2 + 2\delta_1\delta_2 x_1 x_2 + \delta_2^2 x_2^2 + \cdots$$
$$- \epsilon - 2\epsilon\delta_1 x_1 - \epsilon\delta_1^2 x_1^2 - 2\epsilon\delta_2 x_2 + \cdots$$
$$-2 - 2\delta_1 x_1 - 2\delta_2 x_2 + \cdots$$
$$+1$$
$$= 0$$

While the relative magnitude of some of the terms is clear, e.g. $2\delta_1 x_1 \gg \delta_1^2 x_1^2$ and $2\delta_1 x_1 \gg 2\delta_2 x_2$ because $1 \gg \delta_1$ and $\delta_1 \gg \delta_2$ respectively, there is considerable uncertainty about the ordering of other terms, e.g. between $\delta_1^2 x_1^2$ and $2\delta_2 x_2$. Removing the cancelling terms one is left with

$$\delta_1^2 x_1^2 + 2\delta_1\delta_2 x_1 x_2 + \delta_2^2 x_2^2 + \cdots$$
$$-\epsilon - 2\epsilon\delta_1 x_1 - \epsilon\delta_1^2 x_1^2 - 2\epsilon\delta_2 x_2 + \cdots = 0$$

Using $1 \gg \delta_1 \gg \delta_2$ one can see that the leading order terms from the two lines are $\delta_1^2 x_1^2$ and $-\epsilon$. Therefore there are three possible leading order balances:

either	$\delta_1^2 x_1^2$	$= 0$	if	$\delta_1^2 \gg \epsilon$
or	$\delta_1^2 x_1^2 - \epsilon = 0$		if	$\delta_1^2 = \epsilon$
or	$-\epsilon = 0$		if	$\delta_1^2 \ll \epsilon$

Clearly the last option is unacceptable and so too is the first because we require $x_1 = \mathrm{ord}(1)$. Hence we conclude that

$$\delta_1 = \epsilon^{\frac{1}{2}} \quad \text{and} \quad x_1 = \pm 1$$

Removing these two balancing terms leaves as leading order terms $2\delta_1\delta_2 x_1 x_2$ and $-2\epsilon\delta_1 x_1$. Repeating the above arguments

either	$2\epsilon^{\frac{1}{2}}\delta_2 x_1 x_2$	$= 0$	if	$\delta_2 \gg \epsilon$
or	$2\epsilon^{\frac{1}{2}}\delta_2 x_1 x_2 - 2\epsilon^{\frac{3}{2}} x_1 = 0$		if	$\delta_2 = \epsilon$
or	$-2\epsilon^{\frac{3}{2}} x_1 = 0$		if	$\delta_2 \ll \epsilon$

The only acceptable option is

$$\delta_2 = \epsilon \quad \text{and} \quad x_2 = 1 \quad \text{(for both } x_1 \text{ roots)}$$

Because the above determination of the expansion sequence involves some messy intermediate details, in practice one would take two attempts at the problem to determine δ_1 and δ_2. First one would substitute $x = 1 + \delta_1 x_1$ and find $\delta_1 = \epsilon^{\frac{1}{2}}$. Then one would substitute $x = 1 + \epsilon^{\frac{1}{2}} x_1 + \delta_2 x_2$ and find $\delta_2 = \epsilon$. Splitting the problem up into stages, one has to consider at each stage less terms of undetermined magnitude.

Iterative method

Finally the superiority of the iterative method should be noted in cases where the expansion sequence is not known. A suitable rearrangement of the original quadratic is

$$(x - 1)^2 = \epsilon x^2$$

which leads to the iterative process

$$x_{n+1} = 1 \pm \epsilon^{\frac{1}{2}} x_n$$

Starting with $x_0 = 1$, the positive root gives

$$x_1 = 1 + \epsilon^{\frac{1}{2}}$$

and

$$x_2 = 1 + \epsilon^{\frac{1}{2}} + \epsilon$$

Not only is this considerably quicker but there is also no awkward step like the $\epsilon^{\frac{1}{2}}$ level in the expansion method which leaves $x_{\frac{1}{2}}$ undetermined.

Exercise 1.3. Find the first two terms of $x(\epsilon)$ the solution near 0 of

$$\sqrt{2} \sin \left(x + \frac{\pi}{4} \right) - 1 - x + \tfrac{1}{2} x^2 = -\tfrac{1}{6}\epsilon$$

Exercise 1.4. Find the first two terms for all four roots of

$$\epsilon x^4 - x^2 - x + 2 = 0$$

Exercise 1.5 (Stone). Find the first two terms for all three roots of

a: $\epsilon x^3 + x^2 + (2 + \epsilon)x + 1 = 0$
b: $\epsilon x^3 + x^2 + (2 - \epsilon)x + 1 = 0$

1.4   Logarithms

In this section we shall find the solution as $\epsilon \to 0$ (through positive values) of the transcendental equation

$$xe^{-x} = \epsilon$$

One root is near $x = \epsilon$ which is easy to obtain. The other root becomes large as $\epsilon \to 0$ and is more difficult to find. We concentrate on this large root. As the expansion sequence is distinctly unclear, we employ the iterative method.

First we observe that if ϵ is small ($\epsilon < \frac{1}{4}$ is sufficient) the root must lie between $x = \ln(1/\epsilon)$ (for which $xe^{-x} = \epsilon \ln(1/\epsilon) > \epsilon$) and $x = 2\ln(1/\epsilon)$ (for which $xe^{-x} = \epsilon^2 2\ln(1/\epsilon) < \epsilon$). Over this range of x, the x factor merely doubles while the e^{-x} factor falls by an order of magnitude from ϵ to ϵ^2. Thus we can view the x factor as varying weakly and concentrate on the rapid variation in the e^{-x} factor. This suggests the rearrangement of the original equation

$$e^{-x} = \frac{\epsilon}{x}$$

leading to the iterative scheme

$$x_{n+1} = \ln(1/\epsilon) + \ln x_n$$

Further, from the above observations it is clear that the root must lie quite near $\ln(1/\epsilon)$ when ϵ is small. Thus we start the iteration from

$$x_0 = \ln(1/\epsilon)$$

Then

$$x_1 = \ln(1/\epsilon) + \ln\ln(1/\epsilon) = L_1 + L_2$$

where we have introduced the shorthand notation

$$L_1 = \ln(1/\epsilon) \quad \text{and} \quad L_2 = \ln\ln(1/\epsilon)$$

Iterating again

$$x_2 = L_1 + \ln\left[L_1\left(1 + \frac{L_2}{L_1}\right)\right]$$

$$= L_1 + L_2 + \frac{L_2}{L_1} - \frac{L_2^2}{2L_1^2} + \cdots$$

And again

$$x_3 = L_1 + \ln\left[L_1\left(1 + \frac{L_2}{L_1} + \frac{L_2}{L_1^2} - \frac{L_2^2}{2L_1^3}\right)\right]$$

$$= \quad L_1 + L_2 + \left(\frac{L_2}{L_1} + \frac{L_2}{L_1^2} - \frac{L_2^2}{2L_1^3} \right)$$

$$- \tfrac{1}{2} \left(\frac{L_2}{L_1} + \frac{L_2}{L_1^2} + \cdots \right)^2 + \tfrac{1}{3} \left(\frac{L_2}{L_1} + \cdots \right)^3 + \cdots$$

$$= \quad L_1 + L_2 + \frac{L_2}{L_1} + \frac{-\tfrac{1}{2}L_2^2 + L_2}{L_1^2} + \frac{\tfrac{1}{3}L_2^3 - \tfrac{3}{2}L_2^2 + \cdots}{L_1^3} + \cdots$$

The expansion sequence needed by the expansion method is clearly a tough one to guess. Moreover the iterative method produces more than one extra term from each iteration.

The appearance of $\ln \ln(1/\epsilon)$ means that remarkably small values of ϵ are required to achieve a good numerical accuracy of the approximate expressions. Usually one hopes for a tolerable agreement with $\epsilon < 0.5$ or at worse $\epsilon < 0.1$. In order for $\ln \ln(1/\epsilon) > 3$ however one needs $\epsilon < 10^{-9}$. The table below gives the percentage errors at various ϵ for the first five approximations to the large root of our transcendental equation.

ϵ	L_1	$+L_2$	$+L_2/L_1$	$-\tfrac{1}{2}L_2^2/L_1^2$	$+L_2/L_1^2$
10^{-1}	36	12	2	4	0.03
10^{-3}	24	3	0.02	0.04	0.04
10^{-5}	19	1	0.04	0.1	0.001

The table shows that acceptable accuracy is only achieved with many terms of the approximation or with extremely small values of ϵ. The table also demonstrates another common feature of expansions which involve $\ln \ln(1/\epsilon)$. This is that it is unwise to split $(-\tfrac{1}{2}L_2^2 + L_2)/L_1^2$ into two terms, because the error is made worse by the first part before it is eventually improved by the addition of the second part (at least at values of ϵ not astronomically small).

Exercise 1.6. Find several terms in an approximation for the solution of

$$\frac{e^{-x^2}}{x} = \epsilon$$

1.5   Convergence

The expansion method offers little opportunity of proving that an approximation converges. In straightforward problems the form of the nth term will be clear, e.g. ϵ^n, and so one can be satisfied that the expansion is consistent. Just occasionally one can write down the problem for the general n^{th} term, find a strong bound on the magnitude of the term, and thence prove convergence of the expansion. In more difficult problems, however, the expansion sequence will not be clear and one would have no idea of the form of the general term. In these problems one can only be satisfied that the expansion is consistent as far as one has proceeded.

The iterative method on the other hand provides a simple proof of convergence. Suppose $x = x_*$ is the root of the equation

$$x = f(x)$$

where f is used in an iterative process $x_{n+1} = f(x_n)$. Then one iteration will take $x = x_* + \delta$ to

$$f(x_* + \delta) = x_* + \delta f'(x_*) + o(\delta)$$

if δ is small. Thus one iteration will decrease the error if

$$|f'(x_*)| < 1$$

Hence by the contraction mapping theorem, the iterative process will converge onto the root x_* if $|f'(x_*)| < 1$ and if the iteration is started sufficiently near to the root. (The standard theorem needs a small modification to take account of the truncation of the higher order terms which are known to be incorrect after insufficient iterations.)

In the previous sections we had iterative schemes which converge.

In §1.1 $f = \sqrt{1 - \epsilon x}$ with $x_* \sim 1$ so $f'(x_*) \sim -\tfrac{1}{2}\epsilon$

In §1.2 $f = -1/\epsilon + 1/\epsilon x$ with $x_* \sim -1/\epsilon$ so $f'(x_*) \sim -\epsilon$

In §1.3 $f = 1 + \epsilon^{1/2} x$ with $x_* \sim 1$ so $f'(x_*) \sim \epsilon^{1/2}$

In §1.4 $f = \ln(\tfrac{1}{\epsilon}) + \ln(x)$ with $x_* \sim \ln(\tfrac{1}{\epsilon})$ so $f'(x_*) \sim 1/\ln(\tfrac{1}{\epsilon})$

The negative sign of f' in the first two cases means that the error changes sign and so two successive iterations must bracket the answer. Also from the magnitude of f' one can work out how many terms will be correct after a given number of iterations.

1.6   Eigenvalue problems

In this section we consider the eigenvalue problem for the eigenvalue λ associated with the eigenvector $\mathbf{x}$

$$A\mathbf{x} + \epsilon \mathbf{B}(\mathbf{x}) = \lambda \mathbf{x}$$

In order for this to qualify as an algebraic equation, A ought to be a matrix. The techniques of this section, however, can be applied to any linear operator A with adequate compactness. As $\epsilon \mathbf{B}(\mathbf{x})$ is a small perturbation, there is no need for $\mathbf{B}(\mathbf{x})$ to be linear.

We look for the perturbed eigensolution near to a given unperturbed eigensolution with eigenvalue a and associated eigenvector $\mathbf{e}$:

$$A\mathbf{e} = a\mathbf{e}$$

If the matrix A is not symmetric, its transpose will have a different eigenvector $\mathbf{e}^\dagger$ associated with the same eigenvalue:

$$\mathbf{e}^\dagger A = a\mathbf{e}^\dagger$$

Initially we restrict attention to the case where a is a single root with only one independent eigenvector $\mathbf{e}$. Then $\mathbf{e}^\dagger$ is orthogonal to all the other eigenvectors of A.

In the standard way we pose an expansion in powers of ϵ starting from the unperturbed eigensolution

$$\begin{aligned}
\mathbf{x}(\epsilon) &= \mathbf{e} + \epsilon \mathbf{x}_1 + \epsilon^2 \mathbf{x}_2 + \cdots \\
\lambda(\epsilon) &= a + \epsilon \lambda_1 + \epsilon^2 \lambda_2 + \cdots
\end{aligned}$$

Substituting into the governing equation and comparing coefficients of ϵ^n produces

at ϵ^0: $A\mathbf{e} = a\mathbf{e}$ which is automatically satisfied

at ϵ^1: $A\mathbf{x}_1 + \mathbf{B}(\mathbf{e}) = a\mathbf{x}_1 + \lambda_1 \mathbf{e}$

It is useful to rearrange the last equation as

$$(A - a)\mathbf{x}_1 = \lambda_1 \mathbf{e} - \mathbf{B}(\mathbf{e})$$

Now the left hand side of this equation can have no component in the direction of $\mathbf{e}$, because for all $\mathbf{x}_1$

$$\mathbf{e}^\dagger \cdot [(A - a)\mathbf{x}_1] = [\mathbf{e}^\dagger (A - a)] \cdot \mathbf{x}_1 = (a - a)\mathbf{e}^\dagger \cdot \mathbf{x}_1 = 0$$

using the eigenvector property of $\mathbf{e}^\dagger$. Thus there can exist no solution of the equation for $\mathbf{x}_1$ unless the right hand side of the equation also has no component in the direction of **e**, i.e.

$$\mathbf{e}^\dagger \cdot [\lambda_1 \mathbf{e} - \mathbf{B}(\mathbf{e})] = 0$$

Thus we have found the first perturbation of the eigenvalue

$$\lambda_1 = \frac{\mathbf{e}^\dagger \cdot \mathbf{B}(\mathbf{e})}{\mathbf{e}^\dagger \cdot \mathbf{e}}$$

Note that this expression shows that if $\mathbf{B}(\mathbf{e})$ is nonlinear then the eigenvalue is not independent of the magnitude of the eigenvector.

We now can return to the equation for $\mathbf{x}_1$ and substitute the expression for λ_1 to yield

$$(A - a)\mathbf{x}_1 = -\mathbf{B}(\mathbf{e}) + \frac{\mathbf{e}^\dagger \cdot \mathbf{B}(\mathbf{e})}{\mathbf{e}^\dagger \cdot \mathbf{e}}\mathbf{e} = -\mathbf{B}(\mathbf{e})_\perp$$

where the notation $\mathbf{B}(\mathbf{e})_\perp$ has been introduced for that part of $\mathbf{B}(\mathbf{e})$ perpendicular to **e**. Now that the right hand side has no component in the direction of **e** it is possible to invert $(A - a)$ to obtain a solution for $\mathbf{x}_1$, although this solution is not unique because it is possible to add an arbitrary multiple of **e** to $\mathbf{x}_1$ without changing $(A - a)\mathbf{x}_1$. Thus with k_1 an arbitrary scalar

$$\mathbf{x}_1 = -(A - a)^{-1}\mathbf{B}(\mathbf{e})_\perp + k_1\mathbf{e}$$

where the restricted inverse $(A - a)^{-1}$ does exist in the space orthogonal to **e**. If the complete eigendecomposition of A is known, then $\mathbf{x}_1$ can be represented as a sum over all the other eigenvectors $\mathbf{e}^{(j)}$

$$\mathbf{x}_1 = \sum_j{}' \frac{\mathbf{e}^{(j)\dagger} \cdot \mathbf{B}(\mathbf{e})}{(a - a^{(j)})(\mathbf{e}^{(j)\dagger} \cdot \mathbf{e}^{(j)})}\mathbf{e}^{(j)} + k_1\mathbf{e}$$

This completes the first order perturbation.

Second order perturbation

The second order perturbation is governed by

$$A\mathbf{x}_2 + \mathbf{B}_1 = a\mathbf{x}_2 + \lambda_1\mathbf{x}_1 + \lambda_2\mathbf{e}$$

Here $\epsilon\mathbf{B}_1$ is the $\mathrm{ord}(\epsilon)$ change from $\mathbf{B}(\mathbf{e})$ to $\mathbf{B}(\mathbf{e} + \epsilon\mathbf{x}_1)$. If $\mathbf{B}$ is linear, then $\mathbf{B}_1 = B\mathbf{x}_1$. If $\mathbf{B}$ is nonlinear, then $\mathbf{B}_1 = \mathbf{x}_1 \cdot \mathbf{B}'(\mathbf{e})$ where $\mathbf{B}'$ is the first derivative of $\mathbf{B}$. Rearranging the equation for $\mathbf{x}_2$ we have

$$(A - a)\mathbf{x}_2 = \lambda_2\mathbf{e} + \lambda_1\mathbf{x}_1 - \mathbf{B}_1$$

As in the problem for $\mathbf{x}_1$, we must require that the right hand side has no component in the direction of **e**. This leads to the second perturbation in the eigenvalue

$$\lambda_2 = \frac{\mathbf{e}^\dagger \cdot (\mathbf{B}_1 - \lambda_1 \mathbf{x}_1)}{\mathbf{e}^\dagger \cdot \mathbf{e}}$$

and thence, with k_2 an arbitrary scalar, the second perturbation in the eigenvector is

$$\mathbf{x}_2 = -(A - a)^{-1}(\mathbf{B}_1 - \lambda_1 \mathbf{x}_1)_\perp + k_2 \mathbf{e}$$

We see in the expressions for λ_2 and $\mathbf{x}_2$ that it would have been convenient to remove the non-uniqueness in $\mathbf{x}_1$ by requiring it to have no component in the direction of the unperturbed eigenvector, i.e. $\mathbf{e}^\dagger \cdot \mathbf{x}_1 = 0$. In some problems, however, there are more pressing claims than this convenient normalisation.

If the complete eigendecomposition of A is known, and also if $\mathbf{B}$ is linear, then our result for λ_2 can be written in the more familiar form

$$\lambda_2 = {\sum_j}' \frac{(\mathbf{e}^\dagger \cdot B\mathbf{e}^{(j)})(\mathbf{e}^{(j)\dagger} \cdot B\mathbf{e})}{(a - a^{(j)})(\mathbf{e}^{(j)\dagger} \cdot \mathbf{e}^{(j)})(\mathbf{e}^\dagger \cdot \mathbf{e})}$$

Multiple roots

Suppose that the eigenvalue a of A is associated with more than one independent, non-degenerate eigenvector, $\mathbf{e}_1, \mathbf{e}_2, \ldots, \mathbf{e}_n$. We must now consider perturbing around a general eigenvector in the eigenspace

$$\mathbf{x} = \sum_{i=1}^n \alpha_i \mathbf{e}_i + \epsilon \mathbf{x}_1 + \cdots$$
$$\lambda = a + \epsilon \lambda_1 + \cdots$$

Substituting into the governing equation and comparing coefficients of ϵ^n produces at ϵ^1

$$(A - a)\mathbf{x}_1 = \lambda_1 \sum_{i=1}^n \alpha_i \mathbf{e}_i - \mathbf{B}\left(\sum_{i=1}^n \alpha_i \mathbf{e}_i\right)$$

In this case of multiple roots, the left hand side can have no component in the eigenspace. Thus requiring the right hand side to have no component in each of the independent directions $\mathbf{e}_i$ produces

$$\lambda_1 \alpha_1 = \mathbf{e}_1^\dagger \cdot \mathbf{B}\left(\sum_i \alpha_i \mathbf{e}_i\right)/(\mathbf{e}_1^\dagger \cdot \mathbf{e}_1)$$
$$\vdots \qquad \vdots$$
$$\lambda_1 \alpha_n = \mathbf{e}_n^\dagger \cdot \mathbf{B}\left(\sum_i \alpha_i \mathbf{e}_i\right)/(\mathbf{e}_n^\dagger \cdot \mathbf{e}_n)$$

These equations are a new eigenvalue problem in the eigenspace of A to find the eigenvalue λ_1 and eigenvector α. If $\mathbf{B}$ is linear there will exist n eigenvalues and, except in some degenerate cases, n associated independent eigenvectors. If $\mathbf{B}$ is nonlinear, it is possible that no eigensolutions

exist. In such cases the original eigenproblem will have no perturbed
eigensolutions near the eigenspace of the unperturbed problem.

Degenerate roots

Degenerate multiple roots can lead to an expansion in non-integral powers of ϵ. Consider the n-degenerate eigensolution in the Jordan Normal
Form

$$
\begin{aligned}
A\mathbf{e}_1 &= a\mathbf{e}_1 \\
A\mathbf{e}_2 &= a\mathbf{e}_2 + c_2\mathbf{e}_1 \\
&\vdots \\
A\mathbf{e}_n &= a\mathbf{e}_n + c_n\mathbf{e}_{n-1}
\end{aligned}
$$

Then if the perturbation $\epsilon\mathbf{B}(\mathbf{e}_1)$ has a component in the direction $\mathbf{e}_n$, say
ϵB_n, an expansion is needed in powers of $\epsilon^{1/n}$, i.e.

$$
\begin{aligned}
\mathbf{x}(\epsilon) &= \mathbf{e}_1 + \epsilon^{1/n}x_2\mathbf{e}_2 + \epsilon^{2/n}x_3\mathbf{e}_3 + \cdots + \epsilon^{(n-1)/n}x_n\mathbf{e}_n + \cdots \\
\lambda(\epsilon) &= a + \epsilon^{1/n}\lambda_1 + \cdots
\end{aligned}
$$

with solution

$$
x_2 = \lambda_1/c_2, \quad x_3 = \lambda_1^2/c_2c_3, \quad \cdots \quad x_n = \lambda_1^{n-1}/c_2c_3\ldots c_n
$$
$$
\text{and} \quad \lambda_1 = (c_2c_3\ldots c_nB_n)^{1/n}
$$

If the components of $\epsilon\mathbf{B}(\mathbf{e}_1)$ vanish in the directions of $\mathbf{e}_{k+1}, \mathbf{e}_{k+2}, \cdots, \mathbf{e}_n$
then an expansion is needed in powers of $\epsilon^{1/k}$.

Exercise 1.7. Find the second order perturbations of the eigenvalues
of the matrix

$$
\begin{pmatrix} E_1 & 0 \\ 0 & E_2 \end{pmatrix} + \begin{pmatrix} 0 & \omega \\ -\omega & 0 \end{pmatrix}
$$

for small ω and for large ω. Consider to first order the 3×3 version of
this problem.

Exercise 1.8. Find the first order perturbations of the eigenvalues of
the differential equation

$$
y'' + \lambda y + \epsilon y^n = 0
$$

in $0 < x < \pi$, with $y(0) = y(\pi) = 0$ for $n = 1, 2$ and 3.

Asymptotic approximations

2.1 Convergence and asymptoticness

An expansion $\sum^N f_n(z)$ is said to *converge* at a fixed value of z if given an arbitrary $\epsilon > 0$ it is possible to find a number $N_0(z, \epsilon)$ such that

$$\left| \sum_M^N f_n(z) \right| \quad < \quad \epsilon \qquad \text{for all } M, N > N_0$$

Thus an expansion converges if its terms eventually decay sufficiently rapidly.

This property of convergence is less useful in practice than one is usually led to believe. Consider the error function

$$\text{Erf}(z) \quad = \quad \frac{2}{\sqrt{\pi}} \int_0^z e^{-t^2}\, dt$$

Now e^{-t^2} is analytic in the entire complex plane, i.e. it can be expanded in a Taylor series $\sum_0^\infty [(-t^2)^n / n!]$ which converges to the correct value with an infinite radius of convergence. Hence one can integrate term by term to form a series for Erf which also converges with an infinite radius of convergence.

$$\text{Erf}(z) \quad = \quad \frac{2}{\sqrt{\pi}} \sum_0^\infty \frac{(-)^n z^{2n+1}}{(2n+1)\, n!}$$

$$= \quad \frac{2}{\sqrt{\pi}} \left(z - \tfrac{1}{3} z^3 + \tfrac{1}{10} z^5 - \tfrac{1}{42} z^7 + \tfrac{1}{216} z^9 - \tfrac{1}{1320} z^{11} + \cdots \right)$$

Just eight terms of this series will give an accuracy of 10^{-5} up to $z = 1$. As z increases, progressively more terms are needed to maintain this accuracy, e.g. 16 terms to $z = 2$, 31 terms to $z = 3$ and 75 terms to $z = 5$. As well as requiring many terms, some of the intermediate terms are very large when z is large. Thus a computer working with a round-off error of 10^{-7} cannot give an answer correct to 10^{-4} at $z = 3$, because the largest term is about 214, and at $z = 5$ a largest term of 6.6×10^8 leads the computer to 'converge' on an erroneous answer of several hundred. A converging expansion with the terms eventually decaying is thus seen to be of limited practical value when some of the truncated sums are wildly different from the converged limit.

An alternative expression for Erf at large z can be obtained from

$$\mathrm{Erf}(z) \;=\; 1 - \frac{2}{\sqrt{\pi}} \int_z^\infty e^{-t^2}\, dt$$

Integrating by parts

$$\int_z^\infty e^{-t^2}\, dt \;=\; \int_z^\infty \frac{d(-e^{-t^2})}{2t} \;=\; \frac{e^{-z^2}}{2z} - \int_z^\infty \frac{e^{-t^2}}{2t^2}\, dt$$

and again three more times gives

$$= \frac{e^{-z^2}}{2z}\left(1 - \frac{1}{2z^2} + \frac{1.3}{(2z^2)^2} - \frac{1.3.5}{(2z^2)^3}\right) + R$$

where the remainder can be bounded by

$$|R| \;=\; \int_z^\infty \frac{105\, d(e^{-t^2})}{32 t^9} \;<\; \frac{105}{32 z^9}\int_z^\infty d(e^{-t^2}) \;=\; \frac{105 e^{-z^2}}{32 z^9}$$

using $t^{-9} < z^{-9}$ in the range $t > z$. Thus we have proven that as $z \to \infty$

$$\mathrm{Erf}(z) \;=\; 1 - \frac{e^{-z^2}}{z\sqrt{\pi}}\left(1 - \frac{1}{2z^2} + \frac{1.3}{(2z^2)^2} - \frac{1.3.5}{(2z^2)^3} + O(z^{-8})\right)$$

This alternative expansion for the error function diverges. The truncated series, however, is useful: at $z = 2.5$ three terms give an accuracy of 10^{-5}, while above $z = 3$ only two terms are necessary. Our alternative expansion has the important property that the leading term is roughly correct and further terms are corrections of decreasing size. This property is called asymptoticness.

2.2   Definitions

The sum $\sum^N f_n(\epsilon)$ is said to be an *asymptotic approximation* to $f(\epsilon)$ (or alternatively an asymptotic representation of $f(\epsilon)$) as $\epsilon \to 0$, if for each $M \le N$

$$\frac{f(\epsilon) - \sum^M f_n(\epsilon)}{f_M(\epsilon)} \;\to\; 0 \qquad \text{as } \epsilon \to 0$$

i.e. the remainder is smaller than the last term included once ϵ is sufficiently small. If the sum has this asymptotic property, one writes

$$f(\epsilon) \;\sim\; \sum^N f(\epsilon) \qquad \text{as } \epsilon \to 0$$

The words *asymptotic expansion* are sometimes reserved for the case where one can obtain, at least in principle, an indefinite number of terms $(N = \infty)$, although it should be emphasised that worrying about the higher terms as $N \to \infty$ runs counter to the philosophy of asymptoticness in which the first term is virtually correct as $\epsilon \to 0$.

Often the terms $f_n(\epsilon)$ are powers of ϵ multiplied by some coefficient, i.e. $f \sim \sum^N a_n \epsilon^n$ which is called an *asymptotic power series*. We have seen examples in chapter 1, however, which require fractional powers of ϵ and also other functions of ϵ like $\ln(1/\epsilon)$ and $\ln\ln(1/\epsilon)$. In these cases the asymptotic approximations take the form $f \sim \sum^N a_n \delta_n(\epsilon)$ using an *asymptotic sequence* $\delta_0, \delta_1, \ldots$ which has the property $\delta_{n+1}/\delta_n \to 0$ as $\epsilon \to 0$. Note that in order to keep the $\delta_n(\epsilon)$ single-valued ϵ has to be restricted to some sector of the complex ϵ-plane.

When working with ϵ real and positive, the usual condition, a useful class of functions to use for the asymptotic sequence are Hardy's *logarithmico–exponential functions*, which are those functions obtained by a finite number of applications of the operators $+, -, \times, \div, \exp$ and log with the restriction that all the intermediate quantities are real. This class of functions has the important property that any two members can be ordered, i.e. it is possible to decide of two members f and g whether one is smaller than the other, $f = o(g)$ or $g = o(f)$, or whether they are of the same magnitude $f = \text{ord}(g)$.

Exercise 2.1. Why are $\cos(1/\epsilon)$ and $\sin(1/\epsilon)$ not logarithmico–exponential functions, and why can they not be ordered?

2.3 Uniqueness and manipulation

If a function possesses an asymptotic approximation in terms of an asymptotic sequence, then that approximation is *unique* for that particular sequence: given the existence of an approximation $f(\epsilon) \sim \sum^N a_n \delta_n(\epsilon)$ in terms of a given sequence, the coefficients can be evaluated inductively from

$$a_k = \lim_{\epsilon \to 0} \frac{f(\epsilon) - \sum^{k-1} a_n \delta_n(\epsilon)}{\delta_k(\epsilon)}$$

starting at a_0 and proceeding to a_N.

Note that the uniqueness is for one given asymptotic sequence. Thus a single function can have many asymptotic approximations, each in terms of a different asymptotic sequence. For example

$$\tan(\epsilon) \quad \sim \quad \epsilon + \tfrac{1}{3}\epsilon^3 + \tfrac{2}{15}\epsilon^5$$
$$\sim \quad \sin\epsilon + \tfrac{1}{2}(\sin\epsilon)^3 + \tfrac{3}{8}(\sin\epsilon)^5$$
$$\sim \quad \epsilon\cosh(\sqrt{\tfrac{2}{3}}\epsilon) + \tfrac{31}{270}\left(\epsilon\cosh(\sqrt{\tfrac{2}{3}}\epsilon)\right)^5$$

Note also that the uniqueness is for one given function. Thus many different functions can share the same asymptotic approximation, because they can differ by a quantity smaller than the last term included. Consider

$$\exp(\epsilon) \quad \sim \quad \sum_0^\infty \frac{\epsilon^n}{n!} \qquad \text{as } \epsilon \to 0$$
$$\exp(\epsilon) + \exp(-1/\epsilon) \quad \sim \quad \sum_0^\infty \frac{\epsilon^n}{n!} \qquad \text{as } \epsilon \searrow 0$$

Here $\epsilon \searrow 0$ means ϵ *tends down to zero through positive values*. In this example the two functions have an infinite number of terms the same. Two functions sharing the same asymptotic power series, as above, can only differ by a quantity which is not analytic, because two analytic functions with the same power series are identical.

Asymptotic approximations can be naively *added* (subtracted, multiplied or divided) resulting in the correct asymptotic expression for the sum (difference, product or quotient), perhaps based on an enlarged asymptotic sequence. Note the need to be able to order the new asymptotic sequence.

If appropriate to the limiting processes, one asymptotic approximation can be *substituted* into another, although care is needed if an erroneous result is to be avoided. Typically an error occurs when an insufficiently accurate approximation is used in an exponential. For example consider

$$f(z) = \exp(z^2) \qquad \text{and} \qquad z(\epsilon) = \epsilon^{-1} + \epsilon$$

and the appropriate limits $\epsilon \to 0$ and $z \to \infty$. Then without error

$$f(z(\epsilon)) \;=\; \exp(\epsilon^{-2} + 2 + \epsilon^2) \sim \exp(\epsilon^{-2})\, e^2 \left(1 + \epsilon^2 + \tfrac{1}{2}\epsilon^4 + \cdots\right)$$

The poorer, but asymptotic, approximation $z \sim \epsilon^{-1}$ however produces the erroneous result $\exp(\epsilon^{-2})$ for the leading term for f, which is out by the factor e^2. In order to avoid this error, the exponent must be obtained correct to $O(1)$ and not just to leading order. Finally one must remember that cos and sin are exponentials as far as this potential difficulty is concerned.

Asymptotic approximations can be *integrated* term by term with respect to ϵ resulting in the correct asymptotic expression for the integral. Asymptotic approximations cannot however be *differentiated* in general with safety. The trouble comes in differentiating terms like $\epsilon\cos(1/\epsilon)$ which has a differential with respect to ϵ which is not the expected size $O(1)$ but the much larger $O(\epsilon^{-1})$. Such troublesome terms are not analytic. If $f(\epsilon)$ is analytic in some sector of the complex ϵ-plane, one can differentiate term by term in that sector.

2.4 Why asymptotic?

In §1.5 an iterative process was shown to produce a convergent expansion when a certain derivative was less than unity. We now examine the different conditions necessary to make an expansion asymptotic.

Consider the iterative process based on the equation for $f(\epsilon)$

$$f = g + \epsilon A[f]$$

with $g(\epsilon)$ a given function and $A[f]$ a given operator which acts on the function $f(\epsilon)$. Iterating from $f_0 = g$ yields

$$f = g + \epsilon A + \epsilon^2 A \cdot A' + \epsilon^3 \left(\tfrac{1}{2}AA : A'' + (A \cdot A') \cdot A'\right) + \cdots$$

in which A and its first and second derivatives with respect to f, A' and A'', are evaluated at $f = g(\epsilon)$. The iteration can be continued as far as the necessary derivatives of A exist. After N iterations there will be a remainder

$$R_N(\epsilon) = f(\epsilon) - \sum^N \epsilon^n f_n$$

which will be governed by a problem of the form

$$R_N + \epsilon B_N[R_N] = \epsilon^{N+1} h_N$$

in which the operator B_N and the function h_N can be related to the original A and g. To demonstrate that the expansion is asymptotic one needs to prove that, when ϵ is sufficiently small, it is possible to invert the left hand side operator $1 + \epsilon B_N$ for the particular right hand side $\epsilon^{N+1} h_N$, so that R_N exists, and further one must show that the resulting R_N is smaller than the last term included, i.e. $R_N = o(\epsilon^N)$.

This condition for asymptoticness that $1 + \epsilon B_N$ can be inverted as $\epsilon \to 0$ for the *particular* h_N should be contrasted with the requirement for convergence that $|\epsilon A'| < 1$ for A' evaluated near g, i.e.

$$|\epsilon f \cdot A'| \; < \; |f| \qquad \text{for } \textit{all } f$$

Thus convergence can be lost if the operator A' is unbounded.

An example of an operator which is invertible but unbounded is the differential operator. Consider the differential equation for $f(x)$

$$f \; = \; x^{-1} - f'$$

This equation is of the advertised form $f = g + \epsilon A[f]$ as $x \to \infty$ with $\epsilon \equiv \frac{1}{x}$, $g \equiv \frac{1}{x}$ and $A \equiv x \cdot \frac{d}{dx}$. Iterating one obtains the expansion

$$f \; \sim \; x^{-1} + x^{-2} + 2x^{-3} + 2.3x^{-4} + 2.3.4x^{-5}$$

The problem for the remainder is

$$R_N + R_N' \; = \; (N+1)! \, x^{-N-2}$$

which can be solved

$$R_N \; = \; k e^{-x} + (N+1)! \int_x^\infty \frac{e^{x-t}}{t^{N+2}} \, dt$$

Thus $|R_N| \le |k|e^{-x} + (N+1)! \, x^{-N-2}$, i.e. $R_N = o(x^{-N-1})$ as $x \to \infty$. This proves that the expansion is asymptotic. The factorial in the numerator, however, means that the expansion diverges.

Exercise 2.2.   Find the behaviour as $x \to \infty$ of $f(x)$ which satisfies

$$x f' + f \; = \; x^{-3} + \tfrac{1}{2} f''$$

To find the remainder explicitly (and hence prove asymptoticness), it is useful to know that f is related to the Error function by

$$\mathrm{Erf}(x) \; = \; 1 - \frac{1}{\sqrt{\pi}} \left(\frac{1}{x} + f \right) e^{-x^2}$$

Numerical use of diverging series

We have now seen some examples of expansions which are asymptotic and which fail to converge. As explained earlier in this section they are not uncommon in the solution of differential equations. If an expansion is asymptotic, then the leading term is virtually correct once ϵ is sufficiently small. A practical problem arises therefore if the leading term is not sufficiently accurate or if the function is to be evaluated at a value of ϵ which is not sufficiently small. Adding a few extra terms can help, but there is a limit to the number of terms which can be used if the expansion eventually diverges as $N \to \infty$ at a fixed ϵ. It is obviously not sensible to include extra terms once they stop decreasing in magnitude. See chapter 8 on the improvement of the convergence of series.

2.5 Parametric expansions

So far we have only considered functions of a single variable and their approximation. Such problems occur in solving partial differential equations when finding the far field behaviour, and there the approximations are called co-ordinate expansions.

For much of this text, however, we will be concerned with functions of two (or more) variables $f(x,\epsilon)$ and their asymptotic behaviour when one of the variables, ϵ, is small. Typically $f(x,\epsilon)$ will satisfy a differential equation with respect to x in which ϵ occurs as a non-dimensional number or parameter – hence the name parametric expansion.

For functions of two variables we make the obvious generalisation of the definition in §2.2 by allowing the coefficients a_n to become functions of the non-limiting variable x, i.e.

$$ f(x,\epsilon) \quad \sim \quad \sum\nolimits^{N} a_n(x)\delta_n(\epsilon) \qquad \text{as } \epsilon \to 0 $$

If the approximation is asymptotic as $\epsilon \to 0$ for each fixed x, then it is called variously a *Poincaré, classical* or *straightforward* asymptotic approximation.

One can ask if the above pointwise asymptoticness is uniform in x. It is not uncommon, however, for there to be an awkward double limit such as $x \to 0$ and $\epsilon \to 0$ which requires as x decreases to 0 ever tighter restrictions to be placed on ϵ for it to be sufficiently small, e.g. $\epsilon < x$. In such problems a more general form of approximation is necessary:

$$ f(x,\epsilon) \quad \sim \quad \sum\nolimits^{N} a_n(x,\epsilon)\delta_n(\epsilon) \qquad \text{as } \epsilon \to 0 $$

for example with $a_n(x,\epsilon) = b_n(x/\epsilon)$. While the *uniqueness* theorem of §2.3 extends immediately to asymptotic approximations of the Poincaré type, there is no uniqueness for the more general approximations to functions of two or more variables.

2.6   Stokes phenomenon in the complex plane

An analytic function $f(\epsilon)$ has a power series which converges. Thus an
asymptotic expansion which diverges must involve some non-analyticity,
e.g. an essential singularity, and so must be restricted to a sector of the
complex plane. There is therefore the interesting possibility of a single
function possessing several asymptotic expansions, each restricted to
a different sector of the complex plane. This is called a Stokes phenomenon.

Returning to the error function, we found in §2.1 that

$$\text{Erf}(z) \quad \sim \quad 1 - \frac{\exp(-z^2)}{z\sqrt{\pi}} \qquad \text{as } z \to \infty \text{ with } z \text{ real}$$

Looking now on the complex plane, the contour for the integral
$\int_z^\infty \exp(-t^2)\, dt$ can be deformed to any complex z with no change in the
result so long as the contour is kept in the sector where $\exp(-z^2) \to 0$
as $z \to \infty$. Thus the above result is applicable to the sector $|\arg z| < \frac{\pi}{4}$.
A similar expression for the quarter plane about the negative real axis
follows from the fact that $\text{Erf}(z)$ is an odd function of z:

$$\text{Erf}(z) \quad \sim \quad -1 - \frac{\exp(-z^2)}{z\sqrt{\pi}} \qquad \text{as } z \to \infty \text{ with } \frac{3\pi}{4} < \arg z < \frac{5\pi}{4}$$

An expression for Erf which is valid in the top and bottom quarter
planes, where Erf is very large, can be found by evaluating the original
integral from 0 to z using a method given in the next section. The result
is

$$\text{Erf}(z) \quad \sim \quad -\frac{\exp(-z^2)}{z\sqrt{\pi}} \qquad \text{as } z \to \infty \text{ with } \begin{cases} \frac{\pi}{4} < \arg z < \frac{3\pi}{4} \\ \frac{5\pi}{4} < \arg z < \frac{7\pi}{4} \end{cases}$$

Thus we have different asymptotic approximations to the error function which are applicable to four different sectors of the complex plane.
While Erf is analytic at every finite point, there is a non-analytic essential singularity at infinity. Finally it should be noted that the different
sectors can overlap and do not need to be mutually exclusive as in the
example above.

3

Integrals

3.1 Watson's lemma

Finding the asymptotic behaviour as $T \to \infty$ of an integral of the form

$$\int_0^A e^{-zT} f(z)\, dz$$

is common, especially when transform methods are used to solve differential equations. The key to the calculation is that the major contribution to the integral comes from a small region $z = O(1/T)$, with other places contributing exponentially small terms.

We restrict attention to some sector S of the complex plane, in which we assume that f is analytic (except perhaps at $z = 0$) and has an asymptotic approximation

$$f \sim a_0 z^{\alpha_0} + a_1 z^{\alpha_1} + \cdots + a_n z^{\alpha_n} \quad \text{as } z \to 0 \text{ in } S$$

with $-1 < \alpha_0 < \alpha_1 < \ldots < \alpha_n$. If also f is bounded in S (excluding a neighbourhood of $z = 0$), and if $1/T$ and A are in S with $\mathrm{Re}(A) > 0$, then Watson's lemma says that one can integrate term by term to produce the asymptotic result

$$\int_0^A e^{-zT} f(z)\, dz \quad \sim \quad \sum_0^n a_k T^{-\alpha_k - 1} \Gamma(\alpha_k + 1) \qquad \text{as } T \to \infty$$

Proof: Given an arbitrary $\epsilon > 0$, from the asymptoticness of the approximation to $f(z)$ it is possible to find a $z_0(\epsilon)$ such that

$$\left| f(z) - \sum_{k=0}^n a_k z^{\alpha_k} \right| \quad < \quad \epsilon |z^{\alpha_n}| \qquad \text{for all } z \text{ in } S \text{ with } |z| < |z_0|$$

Moreover the complex argument of z_0 can be chosen to be that of $1/T$ so that z_0 is in S and $T z_0$ is real and positive.

Because f is analytic, the path of the integration can be deformed to go radially from 0 to z_0 and then towards the right from z_0 to A, all within S.

27

From the definition of the gamma function $\Gamma(z)$

$$\int_0^{z_0} z^{\alpha_k} e^{-Tz}\, dz - T^{-\alpha_k-1}\Gamma(\alpha_k+1) = -\int_{z_0}^\infty z^{\alpha_k} e^{-Tz}\, dz$$

Using $|e^{-Tz}| < |e^{-(T-1)z_0}|.|e^{-z}|$ for z to the right of z_0, the last integral can be bounded. Thus

$$\left| \int_0^{z_0} e^{-Tz} f(z)\, dz - \sum_0^n a_k T^{-\alpha_k-1}\Gamma(\alpha_k+1) \right|$$

$$< \quad \epsilon \left| T^{-\alpha_n-1}\Gamma(\alpha_n+1) \right|$$

$$+ \quad \left| e^{-(T-1)z_0} \right| \left| \int_{z_0}^\infty \sum_0^n \left(|a_k z^{\alpha_k}| + \epsilon|z^{\alpha_n}| \right) \left| e^{-z} \right| \, |dz| \right.$$

The integral from z_0 to A can also be bounded simply:

$$\left| \int_{z_0}^A e^{-Tz} f(z)\, dz \right| \quad < \quad F e^{-Tz_0}$$

with F the bound on $|f(z)|$ along the contour.

Finally the exponentially small terms are $o(T^{-\alpha_n-1})$ as $T \to \infty$, and ϵ is arbitrarily small, which proves that the approximation to the integral is asymptotic. ∎

Application and explanation

In §2.1 we found an asymptotic approximation to the complement of the error function

$$\mathrm{Erfc}(z) \quad = \quad \frac{2}{\sqrt{\pi}} \int_x^\infty e^{-t^2}\, dt$$

The substitution $t = x + \tau$ puts the integral into the form for Watson's lemma

$$= \quad \frac{2}{\sqrt{\pi}} e^{-x^2} \int_0^\infty e^{-2x\tau} e^{-\tau^2}\, d\tau$$

For an alternative derivation, we first note that the integrand drops exponentially in a small region $\tau = \mathrm{ord}(1/x)$. To concentrate attention on this region we introduce a *rescaling* as in §1.2,

$$\tau = u/x$$

so that

$$\mathrm{Erfc}(z) \quad = \quad \frac{2e^{-x^2}}{x\sqrt{\pi}} \int_0^\infty e^{-2u} e^{-u^2/x^2}\, du$$

Now the significant contributions to the integral come from $u = \mathrm{ord}(1)$ when $x \to \infty$, so that the second exponential can be expanded

$$= \frac{2e^{-x^2}}{x\sqrt{\pi}} \int_0^\infty e^{-2u} \left(1 - \frac{u^2}{x^2} + \frac{u^4}{2x^4} - \frac{u^6}{6x^6} + \cdots \right) du$$

Evaluating the integral term by term, this is

$$\sim \frac{e^{-x^2}}{x\sqrt{\pi}} \left(1 - \frac{1}{2x^2} + \frac{1.3}{(2x^2)^2} - \frac{1.3.5}{(2x^2)^3} \right)$$

The use of the rescaling exposes clearly the importance of the small region, which was effectively hidden in the standard proof of Watson's lemma. The rescaling also gives an early indication of the magnitude of the correction terms. There is however a lack of rigour in the term-by-term integration, because the range of integration includes $u \geq \mathrm{ord}(x)$ for which the expansion is not asymptotic, in addition to $u = \mathrm{ord}(1)$ which gives all the contributions to the integral up to terms of exponential smallness.

Exercise 3.1. Use the method of rescaling to obtain an asymptotic approximation to the exponential integral as $x \to \infty$,

$$E_n(x) = \int_1^\infty t^{-n} e^{-xt} \, dt$$

3.2 Integration by parts

Integrals can sometimes be integrated by parts repeatedly to generate an asymptotic approximation. We have already seen one example in §2.1 with the complement of the error function. The exponential integral in the above exercise can also be integrated by parts. When this method works it has the advantage that it retains an explicit expression for the remainder which can then be bounded to prove the asymptoticness.

A further example of the application of integrating by parts is provided by Fourier series. If the function $f(x)$ is periodic with period 2π and has a continuous integrable derivative, then the expression for the Fourier amplitudes can be integrated by parts once:

$$f_n = \frac{1}{\pi} \int_0^{2\pi} f(x) \cos nx \, dx$$

$$= \frac{f(x) \sin nx}{n\pi} \Big|_0^{2\pi} - \frac{1}{n\pi} \int_0^{2\pi} f'(x) \sin nx \, dx$$

By the periodicity of f the boundary term vanishes, and by the Riemann–Lebesgue lemma the final integral vanishes as $n \to \infty$, rather than being the more apparent ord(1) estimate. Hence for a periodic function with an integrable derivative

$$f_n = o(1/n) \qquad \text{as } n \to \infty$$

It is instructive to look behind this result. If f has a derivative then for x in the small interval $[x_0 - \frac{\pi}{n}, x_0 + \frac{\pi}{n}]$ the function $f(x)$ is roughly equal to the constant value $f(x_0)$ with a change from this value $O(f'\pi/n)$. Multiplying by $\cos nx$ and integrating over the small interval, the constant part integrates to zero due to the *exact cancellation of the oscillation* while the changing part contributes a term $O(f'\pi^2/n^2)$ from the small interval. Adding together the n intervals yields $f_n = O(f'\pi^2/n)$.

An immediate generalisation of the above result is for a periodic function with an integrable k^{th} derivative $f_n = o(n^{-k})$ as $n \to \infty$.

3.3   Steepest descents

The various methods of stationary phase, Laplace, saddle points and steepest descents are all basically the same when viewed as contour integrals on the complex plane, and all are closely related to Watson's lemma. The problem is to evaluate the asymptotic behaviour of an integral of the form

$$\int_C e^{zf(t)}g(t)\,dt \qquad \text{as real, positive } z \to \infty$$

with f and g analytic functions of t and C a given contour on the complex t-plane, most often from infinity in one direction to infinity in another direction.

Global considerations

We start by attempting to estimate the order of magnitude of the integral. When z is real, positive and large the integrand is largest where the real part of f is largest along C. It is therefore useful to contour the function Re(f) on the complex t-plane to find where it is largest. Now one of the Cauchy–Riemann conditions on the analytic f is

$$\nabla^2 \left(\text{Re}(f) \right) = 0$$

Hence Re(f) can have no maxima or minima (except at singular points or branch points where f is not analytic), and so the gradient of Re(f) can only vanish at saddle points. We are thus led to the canonical picture of figure 3.1 for Re(f). The integration contour C must start and end where Re(f) < 0 in order for an infinite integral to converge. Let zRe(f) attain its maximum on C at t_0 and sustain its maximum over a range Δt_0 about t_0. (At this stage it is not necessary to be precise about the definition of the width of the maximum Δt_0, although it can readily be found as $[-z\text{Re}(f''(t_0))]^{-\frac{1}{2}}$.) The order of magnitude estimate for the integral

$$\int_C e^{zf(t)} g(t)\, dt = O\left(e^{zf(t_0)} g(t_0) \Delta t_0\right)$$

proves to be wildly *wrong*. The trouble comes from the imaginary part of f, whose influence has been ignored so far. Now when z is large, $z\text{Im}(f(t))$ varies rapidly within Δt_0 of t_0, and so the integrand oscillates rapidly. A nearly complete cancellation in the integration, as occurred with the Fourier amplitudes in §3.2, nullifies the initial order of magnitude estimate. That this first estimate is too high can be seen by deforming the contour – see figure 3.1 – from C to C_1 (which does not change the value of the integral if f and g are analytic) to produce the

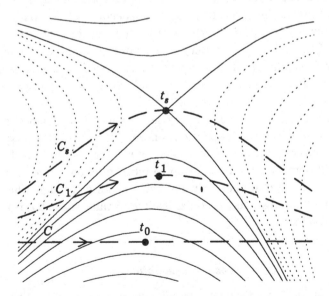

Fig. 3.1 Contours of Re(f) on the t-plane. The continuous curves are for positive values and the dotted curves for negative values.

lower estimate

$$O\left(e^{zf(t_1)}g(t_1)\Delta t_1\right)$$

which is also in error. The integration contour can be pushed further and further down the ridge, producing progressively lower estimates, until the lowest crossing of the ridge is reached at the *saddle point* $t = t_s$, with the estimate

$$O\left(e^{zf(t_s)}g(t_s)\Delta t_s\right)$$

To demonstrate that this estimate is genuine and that the rapid oscillations from the imaginary part no longer cause enormous cancellations, consider the path of the steepest ascent up to the saddle point and the steepest descent away from the saddle. The path is defined to be everywhere parallel to $\nabla \text{Re}(f)$. By a Cauchy–Riemann condition, the path is therefore perpendicular to $\nabla \text{Im}(f)$, i.e. the imaginary part of f, $\text{Im}(f)$, is constant along the path. There is thus strictly no oscillation of the integrand along the path of steepest ascent and descent, and so the order of magnitude estimate is good.

In practice it is not necessary to use the path of steepest ascent and descent: any path descending away from the saddle will produce an integral which converges locally to the correct answer. Analytic methods of evaluating the integral are usually independent of the path, while numerically it is often more expensive to find the steepest path in the vicinity of the point where $\nabla \text{Re}(f) = 0$ than it is to integrate down an arbitrary descending path, which will require a slightly longer range to resolve the slower than steepest decay and also require slightly smaller step sizes to resolve the modest oscillation.

The final global consideration is to decide which is the highest saddle point through which the integration contour must pass, as it progresses across various ridges from one fixed end point to the other. This highest saddle will dominate the integral up to terms of exponential smallness as $z \to \infty$. Note that there can be higher saddles through which the contour does not pass, these saddles leading to valleys isolated from the fixed end points.

Local considerations

We now calculate the leading order contribution to the integral from a saddle point at $t = t_s$. At the saddle $f' = 0$. We assume initially that

$f''(t_s) \neq 0$. Then near to the saddle

$$zf(t) = zf(t_s) + \tfrac{1}{2}z(t - t_s)^2 f''(t_s) + \tfrac{1}{6}z(t - t_s)^3 f'''(t_s) + \cdots$$

Along any descending path the term $\tfrac{1}{2}z(t - t_s)^2 f''(t_s)$ has a negative real part, and when exponentiated it leads to the integrand dropping by a factor of 10^3 before $|t - t_s| > 4|zf''|^{-\frac{1}{2}}$. This suggests a rescaling of the important small region as $z \to \infty$,

$$t = z^{-\frac{1}{2}}\tau$$

Then at fixed τ as $z \to \infty$

$$zf(t) = zf(t_s) + \tfrac{1}{2}\tau^2 f''(t_s) + O(z^{-\frac{1}{2}})$$
$$g(t) = g(t_s) + O(z^{-\frac{1}{2}})$$

The t-integration across the saddle becomes a line integral for τ in a direction with $\mathrm{Re}(\tau^2 f''(t_s)) < 0$. The range of the τ integration is to some large number like $4|f''(t_s)|^{-\frac{1}{2}}$, by which point the integrand has dropped in value by a factor of 10^3. We write this large number as '∞', although it corresponds to a small value of t as $z \to \infty$. Thence

$$\int_{\text{saddle } t_s} e^{zf(t)}g(t)\,dt$$

$$= \int_{`-\infty'}^{`\infty'} e^{zf(t_s)}e^{\frac{1}{2}\tau^2 f''(t_s)}g(t_s)\left(1 + O(z^{-\frac{1}{2}})\right)z^{-\frac{1}{2}}\,d\tau$$

$$= e^{zf(t_s)}g(t_s)\left(\frac{2\pi}{-zf''(t_s)}\right)^{\frac{1}{2}}\left(1 + O(z^{-\frac{1}{2}})\right)$$

Higher order asymptotic approximations can be obtained by retaining the correction terms in $zf(t)$ and $g(t)$, as will be shown in the examples below. Doubts about the justification for the range of the τ-integration can be dispelled by invoking Watson's lemma.

Example : Stirling's formula

Evaluating the integral representation of the factorial function

$$z! = \int_0^\infty t^z e^{-t}\,dt = \int_0^\infty e^{z\ln t - t}\,dt$$

as real, positive $z \to \infty$ provides an example of the case of entirely real quantities where the method of steepest descents is usually called Laplace's method. As a function of t, the integrand is largest where $(z\ln t - t)' = 0$, i.e. at $t = z$. The simple maximum of the integrand has a width (the distance over which the integrand drops by a factor of order unity) $[-(z \ln t - t)'']^{-1/2} = z^{1/2}$. This is large as $z \to \infty$, but is small compared with the distance z to the end point of the integration. With the rescaling

$$t = z + z^{\frac{1}{2}}\tau$$

the exponent becomes

$$z \ln t - t = z \ln z - z + z \ln(1 + z^{-\frac{1}{2}}\tau) - z^{\frac{1}{2}}\tau$$

$$= z \ln z - z - \tfrac{1}{2}\tau^2 + \tfrac{1}{3}z^{-\frac{1}{2}}\tau^3 - \tfrac{1}{4}z^{-1}\tau^4 + \cdots$$

Breaking the exponential into a product of exponentials, and expanding the exponential of the small terms for τ fixed as $z \to \infty$

$$z! \ = \ e^{z \ln z - z} \int_{-\infty'}^{'\infty'} e^{-\frac{1}{2}\tau^2}\left(1 + \left(\tfrac{1}{3}z^{-\frac{1}{2}}\tau^3 - \tfrac{1}{4}z^{-1}\tau^4 + \cdots\right)\right.$$

$$\left. + \tfrac{1}{2}\left(\tfrac{1}{3}z^{-\frac{1}{2}}\tau^3 + \cdots\right)^2 + \cdots\right)z^{\frac{1}{2}}\,d\tau$$

$$\sim \ z^{z+\frac{1}{2}}e^{-z}\sqrt{2\pi}\left(1 + \tfrac{1}{12}z^{-1}\right) \qquad \text{as } z \to \infty$$

This asymptotic approximation for large z is remarkably accurate, with a relative error of less than 10^{-3} down even to $z = 1$.

Finally note in this example the minor generalisation of the method to saddle points which move in the complex t-plane as z varies.

Example : Airy function

The first Airy function can be defined by an integral

$$\text{Ai}(z) \ = \ \frac{1}{2\pi i}\int_C e^{tz - \frac{1}{3}t^3}\,dt$$

with the contour C starting from ∞ with $\arg t = -\tfrac{2}{3}\pi$ and ending at ∞ with $\arg t = \tfrac{2}{3}\pi$. First we find the asymptotic behaviour for z real, positive and large. Figure 3.2 shows the contours of the exponent $\text{Re}(tz - \tfrac{1}{2}t^3)$ on the complex t-plane. There are two saddle points at $t = \pm z^{1/2}$, where $tz - \tfrac{1}{3}t^3 = \pm\tfrac{2}{3}z^{3/2}$. It is necessary only to go through the lower, left hand saddle in order to go over the ridge separating the fixed end points of the integration. The width of the peak of the integrand at $t = -z^{1/2}$ on C is $[-(tz - \tfrac{1}{3}t^3)'']^{-1/2} = 2^{-1/2}z^{-1/4}$. With the rescaling

$$t = -z^{1/2} + z^{-1/4}\tau$$

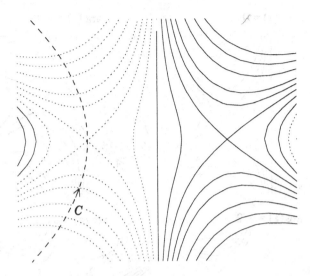

Fig. 3.2   Contours of $\mathrm{Re}(tz - \frac{1}{3}t^3)$. The continuous curves are for positive values and the dotted curves for negative values.

the exponent becomes

$$tz - \tfrac{1}{3}t^3 = -\tfrac{2}{3}z^{3/2} + \tau^2 - \tfrac{1}{3}z^{-3/4}\tau^3$$

Breaking the exponential into a product of exponentials and expanding the exponential of the small terms for τ fixed as $z \to \infty$

$$\mathrm{Ai}(z) = \frac{e^{-\frac{2}{3}z^{3/2}}}{2\pi i} \int_{\text{`}-i\infty\text{'}}^{\text{`}i\infty\text{'}} e^{\tau^2}\left(1 - \tfrac{1}{3}z^{-3/4}\tau^3 + \tfrac{1}{18}z^{-3/2}\tau^6 + \cdots\right)z^{-1/4}\,d\tau$$

$$\sim \frac{e^{-\frac{2}{3}z^{3/2}}}{z^{1/4}2\pi^{1/2}}\left(1 - \tfrac{5}{48}z^{-3/2}\right) \qquad \text{as } z \to \infty$$

When z becomes complex the two saddles move in the complex t-plane, as shown in figure 3.3. To go over the ridge between the two fixed end points of the integration, it is necessary just to go through the same 'left hand' saddle up to $\arg z = \pi$. Thus the above asymptotic approximation is valid in $|\arg z| < \pi$. Note that this saddle is the larger in the range $\frac{1}{3}\pi < \arg z < \pi$. At $\arg z = \pi$ it is necessary to go through both the saddles. As the second is here merely the complex conjugate of the first, no further calculation is needed to obtain

$$\mathrm{Ai}(z) \sim \frac{\sin(\frac{2}{3}(-z)^{3/2} + \frac{\pi}{4})}{(-z)^{1/4}\pi^{1/2}} \qquad \text{for } \arg z = \pi \text{ as } |z| \to \infty$$

Fig. 3.3   The changing contour C for Ai when z is complex.

Beyond arg $z = \pi$ it is necessary to switch to the second saddle, which is higher in the range $\pi < \arg z < \frac{5}{4}\pi$. The last asymptotic approximation for arg $z = \pi$ is in fact asymptotic in the sector $|\arg z - \pi| < \frac{2}{3}\pi$. Thus we have an example of the Stokes phenomenon with different asymptotic approximations applicable to different sectors of the complex z-plane; in this case the sectors overlap.

Exercise 3.2.   Find the asymptotic behaviour of

$$K_\nu(z) \;=\; \tfrac{1}{2} \int_{-\infty}^{\infty} e^{\nu t - z \cosh t} \, dt$$

for real and positive ν and z with $z = \mathrm{ord}(1)$ and $\nu \to \infty$.

Exercise 3.3.   Find the asymptotic behaviour of

$$J_\nu(\nu z) \;=\; \frac{1}{2\pi i} \int_{\infty - i\pi}^{\infty + i\pi} e^{\nu z \sinh t - \nu t} \, dt$$

for real ν and z as $\nu \to \infty$ with first $0 < z < 1$ and second $z = 1$. (The case of $z = 1$ has a cubic saddle where three ridges meet and $f'' = 0$.)

Exercise 3.4.   Find the asymptotic behaviour of

$$P_n(z) \;=\; \frac{1}{2^{n+1}\pi i} \int_C \frac{(t^2 - 1)^n}{(t - z)^{n+1}} \, dt$$

with the contour C enclosing $t = z$, for real z and n with $n \to \infty$, first for $0 < z < 1$ and then for $z > 1$.

3.4   Non-local contributions

The integrals so far have had the form of Watson's lemma in which all the terms in the asymptotic expansion up to those of exponential smallness come from a small region. This need not be the case: at leading order, as in the first example below, or in some higher order correction, as in the second example below, the entire range of the integration can contribute significantly. Terms crucially involving the whole range of integration will be called *global contributions* in contradistinction to *local contributions* from a small region.

Example 1

Consider the simple integral

$$\int_0^1 (\epsilon + x)^{-1/2} \, dx \qquad \text{with exact result} \qquad 2\left((1 + \epsilon)^{1/2} - \epsilon^{1/2}\right)$$

We estimate the contribution to the integral as the magnitude of the function multiplied by the width of the region, for the region near $x = 0$ where $x = \text{ord}(\epsilon)$ and for the majority of the range of integration outside this small region.

If $x = \text{ord}(\epsilon)$, $(\epsilon + x)^{-1/2} = \text{ord}(\epsilon^{-1/2})$ with $\int = \text{ord}(\epsilon^{1/2})$

If $x = \text{ord}(1)$, $(\epsilon + x)^{-1/2} = \text{ord}(1)$ with $\int = \text{ord}(1)$

From these estimates we can conclude that the leading order term is a global contribution, for which the integrand can be approximated by $x^{-1/2}$ and the range of integration is to 1 from a small value outside $x = \text{ord}(\epsilon)$ which we write as '0'. Thus

$$\int_0^1 (\epsilon + x)^{-1/2}\, dx \quad \sim \quad \int_{'0'}^1 x^{-1/2}\, dx \quad = \quad 2$$

To obtain a correction to this leading approximation, we *subtract off* from the original integrand a function whose integral is known exactly and which is equal (or at least asymptotic) to the leading order term. Thus with no approximation

$$\int_0^1 (\epsilon + x)^{-1/2}\, dx \quad = \quad 2 + \int_0^1 \left((\epsilon + x)^{-1/2} - x^{-1/2} \right) dx$$

The contributions to the new integral are now estimated.

If $x = \text{ord}(\epsilon)$,   integrand $= \text{ord}(\epsilon^{-1/2})$ with $\int = \text{ord}(\epsilon^{1/2})$

If $x = \text{ord}(1)$,   integrand $= \text{ord}(\epsilon)$ with $\int = \text{ord}(\epsilon)$

Hence the major contribution comes from the small ϵ region near $x = 0$. To examine the small region it is useful to introduce a *rescaling* $x = \epsilon \xi$, so that $\xi = \text{ord}(1)$ as $\epsilon \to 0$

$$\int_0^1 \left((\epsilon + x)^{-1/2} - x^{-1/2} \right) dx \quad \sim$$

$$\epsilon^{1/2} \int_0^{'\infty'} \left((1 + \xi)^{-1/2} - \xi^{-1/2} \right) d\xi \quad = \quad -2\epsilon^{1/2}$$

It is difficult to proceed to higher terms with further subtractions.

Example 2

Consider the integral

$$\int_0^{\pi/4} \frac{d\theta}{\epsilon^2 + \sin^2 \theta} \quad \text{with exact result} \quad \frac{1}{\epsilon\sqrt{1 + \epsilon^2}} \tan^{-1}\left(\frac{\sqrt{1 + \epsilon^2}}{\epsilon} \right)$$

We estimate the contributions as in the previous example.

If $\theta = \text{ord}(\epsilon)$, integrand $= \text{ord}(\epsilon^{-2})$ with $\int = \text{ord}(\epsilon^{-1})$

If $\theta = \text{ord}(1)$, integrand $= \text{ord}(1)$ with $\int = \text{ord}(1)$

Hence the leading order term is the local contribution, which can be evaluated using the rescaling $\theta = \epsilon u$ (so that the leading contribution comes from $u = \text{ord}(1)$ as $\epsilon \to 0$)

$$\int_0^{\pi/4} \frac{d\theta}{\epsilon^2 + \sin^2 \theta} \quad \sim \quad \int_0^{'\infty'} \frac{\epsilon \, du}{\epsilon^2 + u^2} \quad = \quad \frac{\pi}{2\epsilon}$$

The next term is a global contribution. This can be seen by making a subtraction with

$$\int_0^{\pi/4} \frac{d\theta}{\epsilon^2 + \theta^2} \quad = \quad \frac{1}{\epsilon} \tan^{-1} \left(\frac{\pi}{4\epsilon} \right)$$

Alternatively one can note that the correction term to the integrand in the small region where $\theta = \text{ord}(\epsilon)$,

$$\frac{1}{\epsilon^2 + \sin^2 \theta} \quad = \quad \frac{1}{\epsilon^2 + \epsilon^2 u^2 - \frac{1}{3}\epsilon^4 u^4 + \cdots}$$

$$= \quad \frac{1}{\epsilon^2} \left(\frac{1}{1 + u^2} + \frac{\epsilon^2 u^4}{3(1 + u^2)^2} + \cdots \right)$$

diverges when integrated from $u = 0$ to $u = '\infty'$, with of course '∞' $= \pi/4\epsilon$. The major part of this 'divergence' comes from outside $u = \text{ord}(1)$, which indicates the importance of the whole range of integration to the correction term.

While the method of subtracting off the leading behaviour can be used to obtain one correction term, it becomes cumbersome at higher orders and the alternative method below is easier and more systematic. The idea of subtracting off the singular behaviour of an integral in a small region is, however, frequently used in the numerical evaluation of such integrals, with the subtraction being separately evaluated, usually analytically.

Summing a split range of integration

An alternative method for evaluating integrals with both local and global contributions is to split the range of integration into two at some point $\theta = \delta$ which is large compared with the small region but small compared with the whole interval, i.e. $\epsilon \ll \delta \ll 1$. Asymptotic approximations to the integral over the two ranges are then found separately, using an appropriate rescaling for the small region. When the two parts of the integral are finally combined the result should be independent of the artificially introduced δ, which provides a useful way of detecting slips in algebra.

In order to keep a check on the different errors in the approximations to the integrals over the two ranges, the errors being differently small in the small δ and smaller ϵ, it is useful to tie δ to ϵ as $\epsilon \to 0$, in a way consistent with $\epsilon \ll \delta \ll 1$. This is sometimes called *gearing δ to ϵ*. In this example we consider $\delta = \text{ord}(\epsilon^{1/2})$. Note that it is not necessary to set $\delta = \epsilon^{1/2}$ precisely, and that there are advantages to the self-checking feature in not being so precise.

To approximate the integral over the small range from 0 to δ we use the earlier rescaling $\theta = \epsilon u$ and expand $\sin \theta$ for small θ as $\theta \leq \delta \ll 1$.

$$\int_0^\delta \frac{d\theta}{\epsilon^2 + \sin^2 \theta} \;=\; \frac{1}{\epsilon} \int_0^{\delta/\epsilon} \left(\frac{1}{1 + u^2} + \frac{\epsilon^2 u^4}{3(1 + u^2)^2} + O(\epsilon^4, \epsilon^4 u^2) \right) du$$

The substitution $u = \tan \varphi$ enables one to evaluate the integral exactly

$$=\; \frac{1}{\epsilon} \left[\tan^{-1} \frac{\delta}{\epsilon} + \frac{\epsilon^2}{3} \left(\frac{\delta}{\epsilon} - \frac{3}{2} \tan^{-1} \frac{\delta}{\epsilon} + \frac{1}{2} \frac{\delta/\epsilon}{1 + \delta^2/\epsilon^2} \right) + O(\epsilon \delta^3) \right]$$

We now expand all the terms for large δ/ϵ, using $\tan^{-1} \delta/\epsilon \sim \pi/2 - \epsilon/\delta + \epsilon^3/3\delta^3$. Collecting together terms of a similar order, with $\delta = \text{ord}(\epsilon^{1/2})$,

$$=\; \frac{\pi}{2\epsilon} - \frac{1}{\delta} + 0 + \left(\frac{\epsilon^2}{3\delta^3} + \frac{\delta}{3} \right) - \frac{\pi \epsilon}{4} + O(\epsilon^{3/2})$$

These terms decrease in powers of $\epsilon^{1/2}$ from ϵ^{-1} to $\epsilon^{3/2}$.

To approximate the integral over the remainder of the range from δ to $\pi/4$, we expand $(\epsilon^2 + \sin^2 \theta)^{-1}$ for small ϵ as $\theta \geq \delta \gg \epsilon$.

$$\int_\delta^{\pi/4} \frac{d\theta}{\epsilon^2 + \sin^2 \theta} \;=\; \int_\delta^{\pi/4} \left(\frac{1}{\sin^2 \theta} - \frac{\epsilon^2}{\sin^4 \theta} + O(\frac{\epsilon^4}{\theta^6}) \right) d\theta$$

The integral can be evaluated exactly

$$=\; \cot \delta - 1 - \epsilon^2 \left(\frac{\cos \delta}{3 \sin^3 \delta} + \frac{2 \cos \delta}{3 \sin \delta} - \frac{4}{3} \right) + O\left(\frac{\epsilon^4}{\delta^5} \right)$$

We now expand all the terms for small δ. Collecting together terms of a similar magnitude, with $\delta = \text{ord}(\epsilon^{1/2})$, we have

$$=\; \frac{1}{\delta} - 1 - \left(\frac{\epsilon^2}{3\delta^3} + \frac{\delta}{3} \right) + 0.\epsilon + O(\epsilon^{3/2})$$

These terms decrease in powers of $\epsilon^{1/2}$ from $\epsilon^{-1/2}$ to $\epsilon^{3/2}$.

Bringing together the approximations for the two parts of the range
of the integration, we find that the terms involving δ cancel and so

$$\int_0^{\pi/4} \frac{d\theta}{\epsilon^2 + \sin^2\theta} = \frac{\pi}{2\epsilon} - 1 - \frac{\pi}{4}\epsilon + O(\epsilon^{3/2})$$

Looking back, we can see that the leading order local contribution is
followed by a global first contribution and then a local second correction.

Logarithms

There is a curious intermediate case in which the dominant contribution
is neither truly local nor truly global. Consider evaluating the integral

$$\int_0^\infty f(x,\epsilon)\,dx$$

with a special integrand which has a power law dependence in $\epsilon \ll x \ll 1$
and which maintains the order of magnitude at the ends of this range,
i.e.

$$f(x,\epsilon) \sim \begin{cases} O(\epsilon^{-\alpha}) & \text{in } x = \text{ord}(\epsilon) \\ x^{-\alpha} & \text{in } \epsilon \ll x \ll 1 \\ O(1) & \text{in } x = \text{ord}(1) \end{cases}$$

with $f \to 0$ sufficiently rapidly as $x \to \infty$ for the integral to converge.
We will illustrate the general case with the particular function

$$f(x,\epsilon) = \frac{1}{(\epsilon+x)^\alpha(1+x)}$$

which is precisely of this assumed form. Depending on the value of α
there are three cases for the dominant contribution to the integral.

• If $\alpha < 1$, the integrand increases too slowly as $x \to 0$ for the small
region to be significant. The integral is therefore dominated by $x =$
ord(1), i.e. a global contribution, e.g.

$$\int_0^\infty \frac{dx}{(\epsilon+x)^{1/2}(1+x)} \sim \int_{‘0’}^\infty \frac{dx}{x^{1/2}(1+x)}$$

• If $\alpha > 1$, the integral is dominated by the small region $x = $ ord(ϵ),
i.e. a local contribution, e.g.

$$\int_0^\infty \frac{dx}{(\epsilon+x)^{3/2}(1+x)} \sim \int_0^{‘\infty’} \frac{dx}{(\epsilon+x)^{3/2}}$$

• If $\alpha = 1$, neither $x = $ ord(ϵ) nor $x = $ ord(1) wins, but instead the
dominant contribution comes from the large gap in the orders of magni-

tude between $x = \mathrm{ord}(\epsilon)$ and $\mathrm{ord}(1)$ with a value $\ln(1/\epsilon)$. The two ends contribute slightly smaller $O(1)$ corrections. E.g.

$$\int_0^\infty \frac{dx}{(\epsilon + x)(1 + x)} \;=\; \ln(1/\epsilon) + O(1)$$

(The exact answer is $\ln(1/\epsilon)/(1 - \epsilon)$.) In this case the leading order contribution requires little effort to evaluate. The correction terms from the two ends, however, are only $O(1/\ln(1/\epsilon))$ smaller and so they usually have to be evaluated, unless ϵ is extremely small.

The above results for power law $x^{-\alpha}$ integrands in the range $\epsilon \ll x \ll 1$ hold equally for the slightly more general integrands of the form $x^{-\alpha}(\ln x)^\beta$, with the dominant contribution depending only on whether $\alpha >$ or $=$ or < 1.

Exercise 3.5. Evaluate the first two terms as $r \to 0$ and the first 4 terms (counting $\ln r$ and 1 as different orders of magnitude) as $r \to \infty$ of

$$\int_0^\infty \frac{rx \, dx}{(r^2 + x)^{3/2}(1 + x)}$$

Exercise 3.6. Evaluate the first two terms as $m \nearrow 1$ of the elliptic integral

$$\int_0^{\pi/2} \frac{d\theta}{(1 - m^2 \sin^2 \theta)^{1/2}}$$

Exercise 3.7. Consider the integral

$$\int_0^1 \frac{\ln x}{\epsilon + x} \, dx$$

3.5   An integral equation : the electrical capacity of a long slender body

Integral equations require a combination of the techniques for solving algebraic equations with the techniques for asymptotic evaluation of integrals.

Consider a slender axisymmetric body of length 2 with a surface given in cylindrical polar co-ordinates by

$$r = \epsilon R(z) \qquad \text{for } -1 \le z \le 1$$

with ϵ the slenderness.   A slender ellipsoid of revolution would have $R(z) = \sqrt{1 - z^2}$.

Now from potential theory it is known that the potential φ outside the body can be represented by poles of strength q distributed along the axis within the body, so long as R is continuous with $R = 0$ at the ends and the derivative dR/dz finite. (Unfortunately this restriction excludes cylinders with flat ends.) Thus outside the body

$$\varphi(r, z; \epsilon) \quad = \quad \frac{1}{4\pi} \int_{-1}^{1} \frac{q(z'; \epsilon)\, dz'}{\sqrt{r^2 + (z - z')^2}}$$

To find the electrical capacity of the body, one first has to solve the integral equation to find the function $q(z; \epsilon)$ which gives $\varphi = 1$ on the body surface $r = \epsilon R(z)$ in $|z| \leq 1$. The capacity is then calculated as the total charge

$$\int_{-1}^{1} q(z; \epsilon)\, dz$$

We assume, and check *a posteriori*, that $q(z; \epsilon)$ varies on a length scale with $z = \mathrm{ord}(1)$ and not $\mathrm{ord}(\epsilon)$. Thus we can approximate $q(z'; \epsilon)$ by $q(z; \epsilon)$ for z' nearby z, $|z' - z| \ll 1$. On the other hand if $|z' - z| \gg \epsilon$ we may approximate the distance to a point on the surface $[\epsilon^2 R^2(z) + (z - z')^2]^{1/2}$ by the axial separation $|z' - z|$. Hence

$$\frac{q(z'; \epsilon)}{\sqrt{\epsilon^2 R^2(z) + (z - z')^2}} \quad \sim \quad \frac{q(z; \epsilon)}{|z' - z|} \qquad \text{for } \epsilon \ll |z' - z| \ll 1$$

This integrand is an example of the intermediate case $\alpha = 1$ of the previous section. So long as z is not within $\mathrm{ord}(\epsilon)$ of the ends $z = \pm 1$, there will be two logarithms, one from the z'-integration on each side of z, i.e.

$$\varphi(\epsilon R(z), z) \quad \sim \quad \frac{\ln(1/\epsilon)}{2\pi} q(z; \epsilon) + O(q)$$

The $O(q)$ correction comes from both $|z' - z| = \mathrm{ord}(1)$ and $\mathrm{ord}(\epsilon)$. Imposing the equipotential condition on the surface, $\varphi = 1$, we obtain the leading order approximation to the pole strengths

$$q(z; \epsilon) \quad = \quad \frac{2\pi}{\ln(1/\epsilon)} + O\left(\frac{1}{[\ln(1/\epsilon)]^2}\right)$$

To proceed to higher approximations we pose an expansion for q in powers of $[\ln(1/\epsilon)]^{-1}$ starting from the known leading order,

$$q(z; \epsilon) \quad \sim \quad \frac{q_1(z)}{\ln(1/\epsilon)} + \frac{q_2(z)}{[\ln(1/\epsilon)]^2} + \frac{q_3(z)}{[\ln(1/\epsilon)]^3}$$

with $q_1(z) = 2\pi$. This expansion is substituted into the integral. With q_1 fully determined it is possible to evaluate its integral exactly and hence find the $O(q)$ correction from $|z' - z| = \text{ord}(\epsilon)$ and $\text{ord}(1)$ which forces q_2.

$$\frac{1}{4\pi} \int_{-1}^{1} \frac{2\pi\, dz'}{\sqrt{\epsilon^2 R^2(z) + (z - z')^2}}$$

$$= \; \tfrac{1}{2}\sinh^{-1}\left(\frac{1-z}{\epsilon R(z)}\right) - \tfrac{1}{2}\sinh^{-1}\left(\frac{-1-z}{\epsilon R(z)}\right)$$

$$= \; \ln(1/\epsilon) + \ln\left(\frac{2\sqrt{1-z^2}}{R(z)}\right) + O(\epsilon^2)$$

The $\ln\sqrt{1-z^2}$ comes from $|z - z'| = \text{ord}(1)$, while the $\ln(2/R)$ comes from $|z - z'| = \text{ord}(\epsilon)$. The leading approximation to the integral of q_2 is identical to that for q_1 obtained in the previous paragraph. Combining the approximations for the integrals of q_1 and q_2, we have

$$1 \; = \; \varphi(\epsilon R(z), z)$$

$$= \; 1 + \frac{1}{\ln(1/\epsilon)}\left[\ln\left(\frac{2\sqrt{1-z^2}}{R(z)}\right) + \frac{1}{2\pi}q_2\right] + O\left(\frac{1}{[\ln(1/\epsilon)]^2}\right)$$

Hence

$$q_2(z) = 2\pi \ln\left(\frac{R(z)}{2\sqrt{1-z^2}}\right)$$

To obtain the next correction, q_3, one needs to know the shape $R(z)$. In the case of an ellipsoid of revolution, $R(z) = \sqrt{1-z^2}$, $q_2 = -2\pi\ln 2$ and so one can see that the general term $q_n = 2\pi[-\ln 2]^{n-1}$, which can be summed to $q(z; \epsilon) = 1/\ln(2/\epsilon) + O(\epsilon^2)$!

With our approximation for q we can evaluate the capacity

$$\int_{-1}^{1} q(z; \epsilon)\, dz \; =$$

$$\frac{4\pi}{\ln(1/\epsilon)} + \frac{2\pi}{[\ln(1/\epsilon)]^2} \int_{-1}^{1} \ln\left(\frac{R(z)}{2\sqrt{1-z^2}}\right) dz + O\left(\frac{1}{[\ln(1/\epsilon)]^3}\right)$$

This expression shows the interesting physical phenomenon that the electrical capacity is rather insensitive to the shape, and more important that a sphere with the same electrical capacity would have an enormous volume $\text{ord}([\ln(1/\epsilon)]^{-3})$ compared to the $\text{ord}(\epsilon^2)$ volume of the slender body.

There remains the problem for $q(z)$ within ord(ϵ) of the ends $z = \pm 1$, where the approximation for q calculated above loses its asymptoticness. The trouble near the ends is that there is only one side of z for the z'-integration of $1/|z' - z|$ to produce therefore just one logarithm. Thus one expects q roughly to double in the end region. (To be precise, one needs details of the shape $R(z)$ near the end and also an alternative evaluation of the integral where q is varying rapidly.) This modification of q in the end region leads to a revision of all of its integrals in φ and in the capacitance. Hence there is an $O(\epsilon)$ correction which is negligibly small compared with all the terms in powers of $1/\ln(1/\epsilon)$.

Exercise 3.8. The function $f(t; \epsilon)$ satisfies the integral equation

$$x = \int_{-1}^{1} \frac{f(t; \epsilon)}{\epsilon^2 + (t - x)^2} \, dt \qquad \text{in } -1 \leq x \leq 1$$

Assuming that f remains $O(\epsilon)$ in the end regions where $1 - |t| = \text{ord}(\epsilon)$, obtain the first two terms of an asymptotic approximation for f at fixed $t \neq \pm 1$ as $\epsilon \to 0$. Without any detailed calculation, comment on the contribution to the t-integral from the end regions for a point x outside the end regions, and comment on the behaviour of f in the end regions.

4

Regular perturbation problems in partial differential equations

Progressing from algebraic equations and then integrals, we have now arrived at differential equations. Just a few perturbations of differential equations are regular and so we cover the topic only briefly. In *regular* problems the obvious expansion must be successful, by definition. Partial differential equations can be perturbed in the field equation, in the boundary data and in the position of the boundaries. The last possibility introduces the only novel feature, and so we start with that.

4.1 Potential outside a near sphere

We take a particular near sphere with a surface

$$r \;=\; R(\theta; \epsilon) \;\equiv\; 1 + \epsilon P_2(\cos \theta)$$

where P_2 is a Legendre function. The equation governing the potential φ outside the near sphere is

$$\nabla^2 \varphi \;=\; 0 \qquad \text{in } r \geq R(\theta; \epsilon)$$

$$\text{subject to} \qquad \varphi \;=\; 1 \qquad \text{on } r = R(\theta; \epsilon)$$

$$\text{and} \qquad \varphi \;\rightarrow\; 0 \qquad \text{as } r \rightarrow \infty$$

We pose an expansion for the potential in powers of ϵ,

$$\varphi(r, \theta; \epsilon) \;\sim\; \varphi_0(r, \theta) + \epsilon \varphi_1(r, \theta) + \epsilon^2 \varphi_2(r, \theta)$$

where we expect the leading order φ_0 to be the potential outside an unperturbed sphere ($\epsilon = 0$), $\varphi_0 = 1/r$. Substituting the posed approximation into the field equation and the condition at large r, and then comparing coefficients of ϵ^n, we conclude that each φ_n must be a harmonic function ($\nabla^2 \varphi_n = 0$) which decays at large r.

The problem therefore centres on applying the boundary condition at the perturbed surface. We transfer the condition from the correct

46

boundary $r = R$ to a convenient boundary $r = 1$ by using a Taylor series expansion

$$1 = \varphi\big|_{r=1+\epsilon P_2} = \sum_{n=0}^{\infty} \frac{(\epsilon P_2)^n}{n!} \frac{\partial^n \varphi}{\partial r^n}\bigg|_{r=1}$$

Substituting the expansion for φ and collecting together terms of the same order, we find

$$1 \sim \varphi_0(1,\theta) + \epsilon\left[\varphi_1(1,\theta) + P_2(\cos\theta)\varphi_{0,r}(1,\theta)\right] +$$
$$\epsilon^2\left[\varphi_2(1,\theta) + P_2(\cos\theta)\varphi_{1,r}(1,\theta) + \tfrac{1}{2}P_2^2(\cos\theta)\varphi_{0,rr}(1,\theta)\right]$$

The transferred boundary conditions on the separate φ_n are then found by comparing the coefficients of ϵ^n in this last equation. Thus

at ϵ^0: $\varphi_0(1,\theta) = 1$ and so $\varphi_0 = 1/r$, as expected

at ϵ^1: $\varphi_1(1,\theta) = P_2$ using the known φ_0

The harmonic function which decays at large distances and which satisfies this boundary condition on φ_1 is

$$\varphi_1 = \frac{P_2(\cos\theta)}{r^3}$$

At ϵ^2: $\varphi_2(1,\theta) = 2P_2^2 = \tfrac{36}{35}P_4 + \tfrac{4}{7}P_2 + \tfrac{2}{5}P_0$

using the known φ_0 and φ_1. Hence the harmonic function which decays at large distances and which satisfies the boundary condition on φ_2 is

$$\varphi_2 = \tfrac{36}{35}\frac{P_4(\cos\theta)}{r^5} + \tfrac{4}{7}\frac{P_2(\cos\theta)}{r^3} + \tfrac{2}{5}\frac{1}{r}$$

This last result shows that the electrical capacitance of the near sphere is

$$4\pi\left(1 + \tfrac{2}{5}\epsilon^2 + \cdots\right)$$

Exercise 4.1.   The flow down a slightly corrugated channel is described by a function $w(x,y;\epsilon)$ which is periodic in x and which satisfies

$$\nabla^2 w = -1 \qquad \text{in } |y| \le h(x;\epsilon) \equiv 1 + \epsilon\cos kx$$
$$\text{subject to } w = 0 \qquad \text{on } y = \pm h(x;\epsilon)$$

Obtain the first three terms for w and hence evaluate correct to $\text{ord}(\epsilon^2)$ the average flux per unit width

$$\frac{k}{2\pi}\int_{x=0}^{2\pi/k}\int_{y=-h(x;\epsilon)}^{+h(x;\epsilon)} w(x,y;\epsilon)\,dx\,dy$$

4.2   Deformation of a slowly
rotating self-gravitating liquid mass

This problem combines a perturbation of the position of the boundary with a perturbation of the boundary condition to be applied.

The equilibrium shape of a non-rotating self-gravitating liquid mass is a sphere, say $r = 1$. Let the slow rotation Ω deform the surface slightly to $r = R(\theta; \Omega)$. The gravitational potential $\varphi(r, \theta; \Omega)$ for a liquid of constant density satisfies the equation

$$\nabla^2 \varphi \;=\; \begin{cases} 1 & \text{in } r < R \\ 0 & \text{in } r > R \end{cases}$$

subject to the boundary conditions

$$\varphi \quad \text{and} \quad \partial\varphi/\partial n \quad \text{are continuous at } r = R$$
$$\text{and} \quad \varphi \to 0 \quad \text{as } r \to \infty$$

The liquid mass will be in equilibrium when it is rotating uniformly if its surface is an equipotential for the sum of the gravitational and rotational potentials,

$$\varphi - \tfrac{1}{2}\Omega^2 r^2 \sin^2\theta = \text{a constant, independent of } \theta, \text{ on } r = R(\theta; \Omega)$$

It is also necessary to impose a normalisation condition that the deformed sphere has the same volume as the non-rotating sphere,

$$\int_{r \leq R} dV \;=\; \frac{4\pi}{3}$$

The small parameter Ω only appears in the statement of the problem as Ω^2, so in this regular problem we pose an expansion in powers of Ω^2 (resulting in a deformation which is independent of the direction of rotation)

$$\begin{aligned} \varphi(r, \theta; \Omega) &\sim \varphi_0(r, \theta) &+& \;\; \Omega^2 \varphi_1(r, \theta) \\ R(\theta; \Omega) &\sim \quad 1 &+& \;\; \Omega^2 R_1(\theta) \end{aligned}$$

The boundary conditions on the perturbed surface $r = R$ are transferred back to $r = 1$ using a Taylor series expansion, as in the previous section. In this section we further need an expression for the normal derivative to the deformed surface

$$\frac{\partial\varphi}{\partial n} \;=\; \left[1 + \left(\frac{1}{R}\frac{dR}{d\theta}\right)^2\right]^{-1/2}\left[\frac{\partial\varphi}{\partial r} - \frac{1}{R^2}\frac{dR}{d\theta}\frac{\partial\varphi}{\partial\theta}\right]$$

Actually, accurate to ord(Ω^2), $\partial\varphi/\partial n \sim \partial\varphi/\partial r$. Moreover because φ and $\partial\varphi/\partial n$ are both continuous at $r = R$, one could impose the equivalent boundary condition that φ and $\partial\varphi/\partial r$ be continuous.

Substituting into the governing equations and comparing coefficients of Ω^{2n} generates a sequence of problems.

At Ω^0:
$$\nabla^2\varphi_0 = \begin{cases} 1 & \text{inside,} & \text{`}r < 1\text{'} \\ 0 & \text{outside,} & \text{`}r > 1\text{'} \end{cases}$$
with φ_0 and $\partial\varphi_0/\partial r$ continuous at $r = 1$,
and with φ_0 independent of θ at $r = 1$

Note that the inside is not quite the region $r < 1$, although once the boundary condition has been transferred to $r = 1$ one does tend to work mathematically as if $r < 1$ were the interior. Solving the equations for φ_0 we recover the solution for a non-rotating sphere

$$\varphi_0 = \begin{cases} \frac{1}{6}r^2 - \frac{1}{2} & \text{inside} \\ -\frac{1}{3}r^{-1} & \text{outside} \end{cases}$$

At Ω^2:
$$\nabla^2\varphi_1 = 0 \qquad \text{both inside and outside}$$
$$\varphi_1(1,\theta) + R_1(\theta)\varphi_{0,r}(1,\theta) - \tfrac{1}{2}\sin^2\theta$$
$$\text{is independent of } \theta \text{ and is continuous}$$
$$\varphi_{1,r}(1,\theta) + R_1\varphi_{0,rr}(1,\theta) \qquad \text{is continuous}$$

Now the solution is forced by $\frac{1}{2}\sin^2\theta = \frac{1}{3} - \frac{1}{3}P_2(\cos\theta)$. Thus we try

$$R_1 = aP_2(\cos\theta)$$
$$\varphi_1 = \begin{cases} br^2 P_2(\cos\theta) & \text{inside} \\ cr^{-3}P_2(\cos\theta) & \text{outside} \end{cases}$$

The boundary conditions then give

$$b + \tfrac{1}{3}a + \tfrac{1}{3} = 0 = c + \tfrac{1}{3}a + \tfrac{1}{3} \qquad \text{and} \qquad 2b + \tfrac{1}{3}a = -3c - \tfrac{2}{3}a$$

with solution

$$a = -\tfrac{5}{2} \qquad \text{and} \qquad b = c = \tfrac{1}{2}$$

The negative sign of a means that the equator bulges out and the poles flatten when the liquid mass rotates. Note also that the discontinuity in $\varphi_{1,r}$ at the boundary represents the mass between $r = 1$ and $r = 1 + \Omega^2 R_1$, and thus

$$\nabla^2\varphi_1 = R_1(\theta)\delta(r - 1)$$

Exercise 4.2. The functions $\varphi(x, y; \epsilon)$ and $\lambda(\epsilon)$ satisfy the eigenvalue problem

$$\varphi_{xx} + \varphi_{yy} + \lambda\varphi = 0 \quad \text{in } 0 \le x \le \pi, \quad 0 + \epsilon x(\pi - x) \le y \le \pi$$

subject to $\varphi = 0$ on the boundary. Find the order ϵ correction to φ and λ for the eigensolution

$$\varphi \sim \sin x \sin y \quad \text{and} \quad \lambda \sim 2$$

Exercise 4.3. Slightly nonlinear deep water waves. In a frame moving with the propagation speed c, the surface is steady at $z = \zeta(x)$ and the wave potential $\varphi(x, z)$ satisfies

$$\nabla^2 \varphi = 0 \quad \text{in } z < \zeta(x)$$
$$\varphi \to 0 \quad \text{as } z \to -\infty$$
$$(c + \varphi_x)\zeta_x = \varphi_z \quad \text{on } z = \zeta(x)$$
$$c\varphi_x + \tfrac{1}{2}|\nabla\varphi|^2 + z \quad \text{is independent of } x \text{ on } z = \zeta(x)$$

This is an eigenvalue problem for c. Find the corrections to the linear eigensolution

$$c = 1, \quad \zeta = \epsilon \sin x, \quad \varphi = \epsilon e^z \cos x$$

4.3 Nearly uniform inertial flow past a cylinder

In this final problem the field equation is perturbed. We take the particular nearly uniform flow which has a parabolic perturbation

$$u = 1 + \epsilon y^2 \quad \text{at infinity}$$

In steady, inertially dominated (inviscid) flow the vorticity ω is constant along a streamline. Thus the streamfunction $\psi(x, y; \epsilon)$ is governed by

$$\nabla^2 \psi = -\omega(\psi; \epsilon) \quad \text{in } r > 1$$
$$\text{subject to} \quad \psi = 0 \quad \text{on } r = 1$$
$$\text{and} \quad \psi \to y + \tfrac{1}{3}\epsilon y^3 \quad \text{as } r \to \infty$$

The first problem is to determine the function $\omega(\psi; \epsilon)$. In the far field, we are given $\psi = y + \tfrac{1}{3}\epsilon y^3$, so that there

$$\omega = -\nabla^2 \psi = -2\epsilon y$$

Solving iteratively for y in terms of ψ in the far field, we obtain

$$\omega = -2\epsilon\psi + \tfrac{2}{3}\epsilon^2\psi^3 + O(\epsilon^3\psi^5)$$

We now pose a formal expansion in powers of ϵ,

$$\psi(r, \theta; \epsilon) \sim \psi_0(r, \theta) + \epsilon \psi_1(r, \theta)$$

and substitute into the governing field equation and boundary conditions. Comparing coefficients of ϵ^n we obtain a sequence of problems.

At ϵ^0:
$$\nabla^2 \psi_0 = 0 \qquad \text{in } r > 1$$
$$\text{subject to} \quad \psi_0 = 0 \qquad \text{on } r = 1$$
$$\text{and} \quad \psi_0 \to r \sin \theta \qquad \text{as } r \to \infty$$

with solution

$$\psi_0 = \sin \theta \left(r - \frac{1}{r} \right)$$

At ϵ^1:
$$\nabla^2 \psi_1 = 2\psi_0 = 2 \sin \theta \left(r - \frac{1}{r} \right) \qquad \text{in } r > 1$$
$$\text{subject to} \quad \psi_1 = 0 \qquad \text{on } r = 1$$
$$\text{and} \quad \psi_1 \to \tfrac{1}{3} r^3 \sin^3 \theta \qquad \text{as } r \to \infty$$

This has a solution

$$\psi_1 = \tfrac{1}{3} r^3 \sin^3 \theta - r \ln r \sin \theta - \tfrac{1}{4} \frac{\sin \theta}{r} + \tfrac{1}{12} \frac{\sin 3\theta}{r^3}$$

The presence of the $r \ln r$ term, which grows with r more rapidly than ψ_0, is the first hint that this problem is not really a regular perturbation problem. At the next order it is not possible to find a ψ_2 which behaves satisfactorily at infinity. This problem should in fact be treated like the singular problems of §5.3.

5

Matched Asymptotic Expansions

We now tackle some singular differential equations. There are two distinct types of singular behaviour, which will be studied separately in this chapter and chapter 7. That considered in this chapter typically (but not always – see §5.2) involves a small parameter multiplying the highest derivative. The highest derivative can thus be ignored, so leading to a singular reduction of the order of the equation, except in thin regions of rapid change where the high value of the derivative cancels the effect of the multiplying small parameter. Often these regions of rapid change occur near to the boundary of the domain, and for this reason they are known as boundary layers.

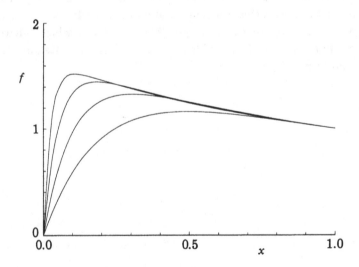

Fig. 5.1   The solution of the problem in §5.1 with $h(x) = e^{-x}$. As ϵ decreases through the values 0.2, 0.1, 0.05 and 0.025 a thin boundary layer of rapid change develops near $x = 0$.

5.1   A linear problem

We start with a simple linear ordinary differential equation, which can be solved exactly. This enables us to probe the structure of the solution and so develop the method of solving asymptotically such singular equations. This model problem is due to Friedrichs. Let $f(x, \epsilon)$ satisfy

$$\epsilon f_{xx} + f_x = h_x \qquad \text{in } 0 < x < 1$$
$$\text{with} \quad f(0, \epsilon) = 0 \qquad \text{and } f(1, \epsilon) = 1$$

where h is a given function with N continuous derivatives on $[0, 1]$. We are interested in the behaviour as ϵ tends to zero through positive values, i.e. $\epsilon \searrow 0$.

5.1.1   The exact solution

By direct integration of the differential equation, we have

$$f(x, \epsilon) \quad = \quad \frac{1}{\epsilon} \int_0^x e^{(t-x)/\epsilon} h(t)\, dt \quad +$$
$$\left(1 - \frac{1}{\epsilon} \int_0^1 e^{(t-1)/\epsilon} h(t)\, dt \right) \frac{1 - e^{-x/\epsilon}}{1 - e^{-1/\epsilon}}$$

We can obtain an asymptotic expansion by integrating by parts M $(< N)$ times,

$$= \quad 1 + \sum_0^{M-1} (-\epsilon)^n \left(h^{(n)}(x) - h^{(n)}(1) \right) + R_I + R_{II}$$

where

$$R_I \quad = \quad - \left(1 + \sum_0^{M-1} (-\epsilon)^n \left(h^{(n)}(0) - h^{(n)}(1) \right) \right) \frac{e^{-x/\epsilon} - e^{-1/\epsilon}}{1 - e^{-1/\epsilon}}$$

$$R_{II} \quad = \quad -(-\epsilon)^{M-1} \int_0^x e^{(t-x)/\epsilon} h^{(M)}(t)\, dt \quad +$$
$$(-\epsilon)^{M-1} \int_0^1 e^{(t-x)/\epsilon} h^{(M)}(t)\, dt \frac{1 - e^{-x/\epsilon}}{1 - e^{-1/\epsilon}}$$

For $\epsilon \searrow 0$ *with $x \neq 0$ fixed*, R_{II} is $O(\epsilon^M)$ by using a bound on $h^{(M)}$, while R_I is exponentially small and so certainly $O(\epsilon^M)$. Thus we have

$$f \quad \sim \quad 1 + \sum_0^{M-1} (-\epsilon)^n \left(h^{(n)}(x) - h^{(n)}(1) \right)$$

an asymptotic series which might diverge if $M = \infty$.

The expansion above is not uniformly asymptotic in x, because the $e^{-x/\epsilon}$ in R_I is not $o(\epsilon^{M-1})$ on the whole interval $0 < x < 1$. Near $x = 0$ we need to have ϵ very small $[< x/((M+)\ln(1/x))]$, i.e. a disaster at $x = 0$. Some singular behaviour of the boundary value problem could have been anticipated, because the equation reduces from second order for $\epsilon > 0$ to first order for $\epsilon = 0$. Thus in the limit we must abandon one boundary condition.

While the above expansion breaks down at $x = 0$, there is an alternative expression which is asymptotic there. For $\epsilon \searrow 0$ *with $\xi = x/\epsilon$ fixed* (ξ is sometimes known as the boundary layer co-ordinate)

$$f(\epsilon\xi, \epsilon) \quad \sim \quad 1 - e^{-\xi} \; +$$
$$\sum_{n=0}^{M-1} (-\epsilon)^n \left(h^{(n)}(0) \left[\sum_{k=0}^{n} \frac{(-\xi)^k}{k!} - e^{-\xi} \right] + h^{(n)}(1) \left(e^{-\xi} - 1 \right) \right)$$

obtained by retaining the appropriate parts of R_I, expanding $h^{(n)}(\epsilon\xi)$ and rearranging. This new asymptotic expansion also breaks down, this time when $\epsilon\xi = \text{ord}(1)$ due to the $\sum \xi^k$.

Thus we see that one function can be represented by two asymptotic expansions. Each expansion is limited by some non-uniformity in the asymptoticness, as $x \to 0$ and $\xi \to \infty$ respectively. The two expansions do, however, take a common form in the common ground where x is small but not too small and where ξ is large but not too large, i.e. $\epsilon \ll x = \epsilon\xi \ll 1$, as they must because they represent the same function.

We now try to obtain the above solution for $f(x, \epsilon)$ by solving exactly some approximate problems, rather than approximating the exact solution.

5.1.2   The outer approximation

We start by naively treating the problem as a regular perturbation problem, even though we know it is not. In this way we produce what is known as the outer approximation, which will often be referred to as the 'outer'. Thus we formally pose a Poincaré expansion with $P < N$,

$$f(x, \epsilon) \quad \sim \quad \sum_{n=0}^{P} \epsilon^n f_n(x)$$

Substituting into the governing equation and boundary conditions and comparing coefficients of ϵ^n yields at ϵ^0: $f_0' = h'$, $f_0(0) = 0$ & $f_0(1) = 1$

at ϵ^n: $f_n' = -f_{n-1}''$, $f_n(0) = 0$ & $f_n(1) = 0$ for $n > 1$

Both boundary conditions cannot be satisfied in general, because the problem is not really regular. In order to satisfy the boundary conditions we will need one or more boundary layers in which the above outer approximation is not appropriate. We postpone until §5.1.7 a discussion of where the boundary layers might be. Here we use our knowledge of the exact solution to decide that there will be a boundary layer at $x = 0$ and none at $x = 1$. Thus the above outer must satisfy just the boundary condition at $x = 1$. Hence

$$
\begin{aligned}
f_0(x) &= h(x) - h(1) + 1 \\
f_n(x) &= (-)^n \left(h^{(n)}(x) - h^{(n)}(1) \right) \qquad \text{for } n > 1
\end{aligned}
$$

A special feature of our model problem is that the outer is now completely determined. Normally one would have some undetermined constants of integration in the outer solution at this stage.

5.1.3   The inner approximation (or boundary layer solution)

The singular reduction of the equation from second order to first order is not appropriate if there are large gradients in thin regions. From the exact solution we know that there is such a thin region near $x = 0$ with a width ϵ. In §5.1.6 we examine how the width of the thin region can be determined. The thin region is studied by introducing a rescaling with a stretched co-ordinate

$$
\xi = x/\epsilon
$$

The governing equation then becomes when $M < N$

$$
\begin{aligned}
\frac{1}{\epsilon} f_{\xi\xi} + \frac{1}{\epsilon} f_\xi &= h_x(\epsilon\xi) \\
&= \sum_{n=1}^{M} \epsilon^{n-1} h^{(n)}(0) \frac{\xi^{n-1}}{(n-1)!} + o(\epsilon^{M-1} \xi^{M-1})
\end{aligned}
$$

We seek a formal expansion of the Poincaré form for $\epsilon \searrow 0$ at fixed ξ with $Q < N$,

$$
f(x, \epsilon) \sim \sum_0^Q \epsilon^n g_n(\xi)
$$

We substitute into the governing equation and boundary condition at $x = 0$. Note that the boundary condition at $x = 1$ cannot be applied, because it is not an accessible point for fixed ξ as $\epsilon \searrow 0$. Comparing coefficients of ϵ^n yields

at ϵ^{-1}: $g_0'' + g_0' = 0$, $g_0(0) = 0$

at ϵ^{-1+n}: $g_n'' + g_n' = h^{(n)}(0)\frac{\xi^{n-1}}{(n-1)!}$, $g_n(0) = 0$ for $n > 1$

with solutions

$$g_0 = A_0(1 - e^{-\xi})$$

$$g_n = A_n(1 - e^{-\xi}) + (-)^n h^{(n)}(0) \sum_{k=1}^{n} \frac{(-\xi)^k}{k!} \qquad \text{for } n > 1$$

with constants of integration A_n. These constants will be determined in the next subsection by applying in some way the other boundary condition at $x = 1$, which is not immediately accessible to this thin inner region.

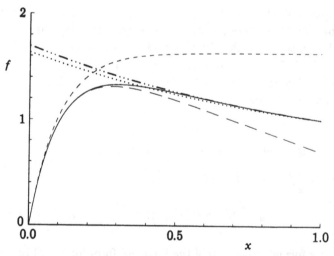

Fig. 5.2   The outer and inner approximations for $h(x) = e^{-x}$ and $\epsilon = 0.1$. The continuous curve is the exact solution. The leading order outer approximation is given by the dotted curve, while the dashed–dotted curve includes the first correction. The leading order inner approximation is given by the short dashed curve, while the long dashed curve includes the first correction.

5.1.4 Matching

We now have two asymptotic expansions for the solution, one for fixed
x and one for fixed ξ. We now will see that these two expansions are of
a similar form in an overlap region which has both x small and ξ large,
i.e. $\epsilon \ll x = \epsilon\xi \ll 1$. Forcing the two expansions to be equal in the
overlap determines the unknowns – here just the A_n. This process is
called matching.

We express both the outer and the inner in terms of an intermediate
variable

$$\eta = x/\epsilon^\alpha = \xi\epsilon^{1-\alpha} \qquad \text{with } 0 < \alpha < 1$$

We then take the limit $\epsilon \searrow 0$ with η fixed, which makes $x \to 0$ and
$\xi \to \infty$. To help organise the terms of differing sizes, it is useful to think
of some particular value for α, say $\frac{1}{2}$. In the overlap region the outer
becomes

$$[h(0) - h(1) + 1] + \epsilon^\alpha \eta h'(0) + \tfrac{1}{2}\epsilon^{2\alpha}\eta^2 h''(0) + \tfrac{1}{6}\epsilon^{3\alpha}\eta^3 h'''(0) + \cdots$$
$$+\epsilon[h'(1) - h'(0)] - \epsilon^{1+\alpha}\eta h''(0) - \epsilon^{1+2\alpha}\eta^2 h'''(0) + \cdots$$
$$+\epsilon^2[h''(0) - h''(1)] + \cdots$$
$$+ \cdots$$

where the successive rows come from the successive f_n. In the overlap
region the inner becomes

$$
\begin{array}{ll}
A_0 & + \quad E.S.T. \\
+\epsilon^\alpha \eta h'(0) + \epsilon A_1 & + \quad E.S.T. \\
+\tfrac{1}{2}\epsilon^{2\alpha}\eta^2 h''(0) - \epsilon^{1+\alpha}\eta h''(0) + \epsilon^2 A_2 & + \quad E.S.T. \\
+ \cdots &
\end{array}
$$

where the successive rows come from the successive g_n, and $E.S.T.$
stands for *Exponentially Small Terms*, e.g. $A_0 \exp(-\eta/\epsilon^{1-\alpha})$. Comparing the two expressions, we see that they are identical if we set

$$
\begin{aligned}
A_0 &= h(0) - h(1) + 1 \\
A_1 &= h'(1) - h'(0) \\
\text{and } A_2 &= h''(0) - h''(1)
\end{aligned}
$$

which fully determines the solution.

Note that some terms have jumped their order, e.g. the term $\epsilon^\alpha \eta h'(0)$
comes out of the ord(ϵ) term ϵg_1. To stop the first ignored term in the
inner $O(\epsilon^{Q+1}\xi^{Q+1})$ being larger than the last retained unknown term
$\epsilon^Q A_Q$, it is necessary to take $\alpha > Q/(Q+1)$. This is permissible because

we only need $0 < \alpha < 1$. Thus in the above displays, $Q = 2$, and so we should have chosen $\alpha > \frac{2}{3}$. However from the structure of the α and η dependencies of the terms, it is clear that nothing has been overlooked by our book-keeping with $\alpha \approx \frac{1}{2}$.

The order jumping term $\epsilon \xi h'(0)$ in ϵg_1 becomes as large as g_0, i.e. the inner loses its asymptoticness, when $\epsilon \xi = \mathrm{ord}(1)$. This tells us that $h(x)$, which was considered a small correction in the equation governing the inner problem, can no longer be considered a small correction at $x = \mathrm{ord}(1)$, i.e. there we need a new balance – the outer.

Exercise 5.1. The function $y(x; \epsilon)$ satisfies

$$\epsilon y'' + (1 + \epsilon)y' + y = 0 \qquad \text{in } 0 \le x \le 1$$

and is subject to boundary conditions $y = 0$ at $x = 0$ and $y = e^{-1}$ at $x = 1$. Find two terms in the outer approximation, applying only the boundary condition at $x = 1$. Next find two terms in the inner approximation for the boundary layer near to $x = 0$, which can be assumed to have a width $\mathrm{ord}(\epsilon)$, and applying only the boundary condition at $x = 0$. Finally determine the constants of integration in the inner approximation by matching.

5.1.5   Van Dyke's matching rule

Matching with an intermediate variable η can be tiresome. Van Dyke's rule usually works and is more convenient.

First we introduce some notation for our two limit operations.

$$\begin{aligned} E_P f &= \text{outer limit } (x \text{ fixed, } \epsilon \searrow 0) \text{ retaining } P + 1 \text{ terms} \\ &= \sum_0^P \epsilon^n f_n(x) \\ H_Q f &= \text{inner limit } (\xi \text{ fixed, } \epsilon \searrow 0) \text{ retaining } Q + 1 \text{ terms} \\ &= \sum_0^Q \epsilon^n g_n(\xi) \end{aligned}$$

With this notation Van Dyke's matching rule is

$$E_P H_Q f = H_Q E_P f$$

In words this means the following. First one takes the inner solution to $Q + 1$ terms and substitutes x/ϵ for ξ. The outer limit of x fixed as

$\epsilon \searrow 0$ is then taken retaining $P+1$ terms. This produces the left hand side of the above equation. A similar process with the inner and outer interchanged produces the right hand side. Requiring that the left and right hand sides of the equation are identical then determines some of the constants of integration.

- For example for $P = Q = 0$

$$
\begin{aligned}
E_0 H_0 f &= E_0 \left\{ A_0 (1 - e^{-\xi}) \right\} \\
&= E_0 \left\{ A_0 (1 - e^{-x/\epsilon}) \right\} \\
&= A_0 \\
H_0 E_0 f &= H_0 \left\{ h(x) - h(1) + 1 \right\} \\
&= H_0 \left\{ h(\epsilon\xi) - h(1) + 1 \right\} \\
&= h(0) - h(1) + 1
\end{aligned}
$$

And so the constant A_0 is determined to be $h(0) - h(1) + 1$.

- For example for $P = Q = 1$

$$
\begin{aligned}
E_1 H_1 f &= E_1 \left\{ A_0 (1 - e^{-\xi}) + \epsilon[A_1 (1 - e^{-\xi}) + h'(0)\xi] \right\} \\
&= E_1 \left\{ A_0 (1 - e^{-x/\epsilon}) + \epsilon[A_1 (1 - e^{-x/\epsilon}) + h'(0)x/\epsilon] \right\} \\
&= A_0 + x h'(0) + \epsilon A_1 \\
H_1 E_1 f &= H_1 \left\{ h(x) - h(1) + 1 - \epsilon[h'(x) - h'(1)] \right\} \\
&= H_1 \left\{ h(\epsilon\xi) - h(1) + 1 - \epsilon[h'(\epsilon\xi) - h'(1)] \right\} \\
&= h(0) - h(1) + 1 + \epsilon\xi h'(0) - \epsilon h'(0) + \epsilon h'(1)
\end{aligned}
$$

Because the $xh'(0)$ in $E_1 H_1 f$ is equal to $\epsilon\xi h'(0)$ in $H_1 E_1 f$, the matching rule is successful and the constants are correctly determined as

$$
A_0 = h(0) - h(1) + 1 \quad \text{and} \quad A_1 = h'(1) - h'(0)
$$

Van Dyke's matching rule does not always work – see §5.2.5. Moreover the rule does not show that the inner and outer are identical in an overlap region.

We have determined the integration constants by matching. A cruder method is to *patch*, in which the value of the inner is set equal to the value of the outer at some particular x. If there is more than one constant to be determined, several x can be used or several derivatives of f at one x. In numerical work it can be difficult to implement a proper matching, and then one uses the unsatisfactory patching method.

Exercise 5.2. Try $P = Q = 2$ and the off diagonal case $P = 0$ with $Q = 1$.

Exercise 5.3. Show that Van Dyke's matching rule will work for general P and Q with the above outer and inner solutions of the particular model equation.

5.1.6   Choice of stretching

In §5.1.3 we used the exact solution to decide the appropriate stretching of the thin region near $x = 0$. In general one has to examine all the possible stretching transformations, as in the rescaling of the singular algebraic equation in §1.2.

Consider the transformation

$$x = \epsilon^{\alpha}\eta$$

This will expand the region near $x = 0$ if $\alpha > 0$. If α is not an integer, the expansion sequence for f will include non-integer powers of ϵ. The iterative method is therefore more appropriate until the expansion sequence becomes clear. Sometimes a more general stretching than ϵ^{α} is needed. In a nonlinear problem, one may also have to stretch the dependent variable, here f, possibly differently in different regions.

Substituting into the governing equation, we have for $M < N$

$$\epsilon^{1-2\alpha} f_{\eta\eta} + \epsilon^{-\alpha} f_{\eta} = h_x(\epsilon^{\alpha}\eta)$$

$$= \sum_{n=0}^{M-1} \epsilon^{n\alpha} h^{(n+1)}(0)\frac{\eta^n}{n!} + o(\epsilon^{\alpha(M-1)})$$

We now scan all possible $\alpha > 0$, starting with large α and then decreasing.

• If $\alpha > 1$, i.e. a finer scaling than the known boundary layer, the above equation can be rearranged placing all the small correction terms on the right hand side of the equation:

$$f_{\eta\eta} = -\epsilon^{\alpha-1} f_{\eta} + \epsilon^{2\alpha-1} h_x(\epsilon^{\alpha}\eta)$$

Solving iteratively, we find

$$
\begin{aligned}
f = {} & A + B\left(\eta - \tfrac{1}{2}\epsilon^{\alpha-1}\eta^2 + \tfrac{1}{6}\epsilon^{2(\alpha-1)}\eta^3 - \tfrac{1}{24}\epsilon^{3(\alpha-1)}\eta^4 + \cdots\right) \\
& + \epsilon^{2\alpha-1} h'(0)\left(\tfrac{1}{2}\eta^2 - \tfrac{1}{6}\epsilon^{\alpha-1}\eta^3 + \tfrac{1}{24}\epsilon^{2(\alpha-1)}\eta^4 + \cdots\right) \\
& + \epsilon^{3\alpha-1} h''(0)\left(\tfrac{1}{6}\eta^3 - \tfrac{1}{24}\epsilon^{\alpha-1}\eta^4 + \cdots\right) \\
& + \cdots
\end{aligned}
$$

Further terms horizontally come from iterating again through the small correction term on the right hand side of the governing equation $-\epsilon^{\alpha-1}f_\eta$, while further rows come from further terms in the expansion of h_x about $x = 0$. The constants A and B are available at every order.

Applying the boundary condition $f = 0$ at $x = 0$ yields $A = 0$ at all orders. The constant(s) B are not determined by this boundary condition.

We now find that the solution cannot be matched to the outer solution by any method. Applying Van Dyke's rule at the leading order, i.e. with $P = Q = 0$, one cannot match the inner's linear term $B\eta$ to the outer's approximately constant term $h(x) - h(1) + 1$ (unless by chance $h(0) - h(1)+1 = 0$). The mismatch is worse at higher orders. When attempting to match in an overlap region with an intermediate variable, it becomes clear that one cannot break through the barrier $\eta \ll \epsilon^{1-\alpha}$, because at $\epsilon^{\alpha-1}\eta = \mathrm{ord}(1)$ an infinite number of terms – all those on a horizontal line – become the same size.

The solution for $\alpha > 1$ is not entirely spurious. It can be matched to our inner solution, because the above expression can be written

$$(A = 0) + B\epsilon^{1-\alpha}\left(1 - e^{-\epsilon^{\alpha-1}\eta}\right) + h'(0)\epsilon\left(e^{-\epsilon^{\alpha-1}\eta} - 1 + \epsilon^{\alpha-1}\eta\right)$$
$$+ h''(0)\epsilon^2\left(-e^{-\epsilon^{\alpha-1}\eta} + 1 - \epsilon^{\alpha-1}\eta + \tfrac{1}{2}\epsilon^{2(\alpha-1)}\eta^2\right) + \cdots$$

which is of course the inner. Hence the solution for $\alpha > 1$ is simply an expansion of the inner on the finer length scale.

• If $\alpha = 1$, then we have the inner scaling of §5.1.3 which we know works.

• If $0 < \alpha < 1$, i.e. a coarser scaling than the known boundary layer, the governing equation can be rearranged placing all the small correction terms on the right hand side of the equation:

$$f_\eta = \epsilon^\alpha h_x(\epsilon^\alpha\eta) - \epsilon^{1-\alpha}f_{\eta\eta}$$

Solving iteratively again, we find

$$\begin{aligned}
f = {} & A + \epsilon^\alpha\eta h'(0) \\
& + \tfrac{1}{2}\epsilon^{2\alpha}\eta^2 h''(0) - \epsilon^{\alpha+1}\eta h''(0) \\
& + \tfrac{1}{6}\epsilon^{3\alpha}\eta^3 h'''(0) - \tfrac{1}{2}\epsilon^{2\alpha+1}\eta^2 h'''(0) + \epsilon^{\alpha+2}\eta h'''(0) \\
& + \cdots
\end{aligned}$$

The terms on each horizontal line come from iterating through the small correction term $\epsilon^{1-\alpha}f_{\eta\eta}$ on the right hand side. Note that this produces only a finite number of terms for each row. Further rows come from further terms in the expansion of h_x about $x = 0$. The constant A is available at every order.

Applying the boundary condition at $x = 0$ yields $A = 0$ at all orders.

We now find that the solution cannot be matched to the outer solution by any method. The leading order terms $\epsilon^\alpha \eta$ and $h(x) - h(1) + 1$ cannot be matched either by Van Dyke's rule or by an intermediate variable in an overlap region.

The solution for $0 < \alpha < 1$ is also not entirely spurious. It can be matched to the outer if we do not take $A = 0$, i.e. we forgo applying the boundary condition which was the very reason for having a boundary layer at $x = 0$. If instead of $A = 0$, we take

$$A \quad \sim \quad [h(0) - h(1) + 1] + \epsilon[h'(1) - h'(0)] + \epsilon^2[h''(0) - h''(1)]$$

then we recover the form of f found earlier in §5.1.4 in the overlap region.

From examining all the possible values of $\alpha > 0$, we conclude that it is only possible to match with the outer if $\alpha \leq 1$, while it is only possible to include the ϵf_{xx} as a main term and so satisfy the boundary condition if $\alpha \geq 1$. As the inner must have the two properties, we conclude that the inner must have the scaling of $\alpha = 1$.

It is interesting to watch the relative importance of the terms in the equation as α varies.

	ϵf_{xx}	$+$	f_x	$=$	h_x	
$\alpha = 0$			......balance......			the outer
$0 < \alpha < 1$			dominant			the overlap
$\alpha = 1$	......balance......					the inner
$1 < \alpha$	dominant					the sub-inner

The reason that the inner expansion (governed by one equation) can be matched to the outer expansion (governed by a different equation at leading order), is that there exists an intermediate expansion in an overlap region. The intermediate expansion is governed by an intermediate equation which at leading order is the common terms of the leading order of the equations for the inner and the outer. It is the lack of any common terms between the sub-inner (with $\alpha > 1$) and the outer which makes it impossible to match them with an intermediate variable.

The potentially interesting scalings of the equation (rescaling both dependent and independent variables in a nonlinear problem) are those which produce a balance between two or more terms in the equation. Such scalings are sometimes called *distinguished limits*.

Exercise 5.4. Find the rescaling of x near $x = 0$ for

$$\epsilon x^m y' + y = 1 \quad \text{in } 0 < x < 1 \quad \text{with } y(0) = 0$$

when $0 < m < 1$

Exercise 5.5. (Stone)   The function $y(x; \epsilon)$ satisfies

$$\epsilon y'' + x^{1/2} y' + y = 0 \quad \text{in } 0 \le x \le 1$$

and is subject to boundary conditions $y = 0$ at $x = 0$ and $y = 1$ at $x = 1$. First find the rescaling for the boundary layer near $x = 0$, and obtain the leading order inner approximation. Then find the leading order outer approximation and match the two approximations.

5.1.7   Where is the boundary layer?

We simplified §5.1.2 by assuming that the boundary condition at $x = 1$ could be applied to the outer and that no boundary layer was needed there. We now see what would have happened if we had tried to put an unnecessary boundary layer near $x = 1$. From this study, we shall learn how to anticipate where boundary layers will be needed.

We expand the region near $x = 1$ by using a stretched co-ordinate $x = 1 - \epsilon^\alpha \eta$, with $\alpha > 0$ for a stretching and $\eta > 0$. The governing equation then becomes

$$\epsilon^{1-2\alpha} f_{\eta\eta} - \epsilon^{-\alpha} f_\eta = h_x(1 - \epsilon^\alpha \eta)$$

As in §5.1.6, the choices $0 < \alpha < 1$ and $\alpha > 1$ do not give useful balances in the equation. So we look at $\alpha = 1$:

$$f_{\eta\eta} - f_\eta = \epsilon \sum_{n=0}^{M-1} \frac{(-\epsilon\eta)^n}{n!} h^{(n+1)}(1) + o(\epsilon^M)$$

with $M < N$.

Solving iteratively, and applying the boundary equation at $x = 1$, i.e. $\eta = 0$,

$$f \sim 1 + A(e^\eta - 1) - \epsilon\eta h'(1) + \epsilon^2 \left(\tfrac{1}{2}\eta^2 + \eta)\right)h''(1)$$

with the constant A available at all orders.

The above inner solution near $x = 1$ has to be matched to the outer solution to which no boundary condition has been applied, i.e.

$$f \sim B + h(x) - \epsilon h'(x) + \epsilon^2 h''(x)$$

with the constant B available at all orders.

Matching by Van Dyke's rule or by the intermediate variable is successful and determines the integration constants

$$A = 0 \quad \text{and} \quad B \sim 1 - h(1) + \epsilon h'(1) - \epsilon^2 h''(1)$$

With $A = 0$, the inner solution becomes merely a re-expression of the outer in terms of the stretched co-ordinate. This is quite different to the boundary layer at $x = 0$ which through terms like $[h(0) - h(1) + 1] e^{-\xi}$ deviates from the outer (re-expressed in terms of the stretched co-ordinate there).

The difficulty in trying to place a boundary layer at $x = 1$ is that the additional solution of the equation, which enters when the stretched co-ordinates restore the order of the governing equation to 2, blows up exponentially away from $x = 1$ (and so must be zero), while it decays away from $x = 0$. To have a non-trivial boundary layer one needs some extra solutions for the inner which decay into the outer region. They do not, however, have to decay exponentially.

- Example 1.   Consider

$$\epsilon^2 f'' - f = -1 \quad \text{in } 0 < x < 1 \quad \text{with} \quad f = 0 \quad \text{at } x = 0 \text{ and } 1$$

For the stretchings ϵ^α with $\alpha = 1$, this equation has exponentially decaying solutions for x increasing and for x decreasing. Thus boundary layers are possible both near $x = 0$ and near $x = 1$, and both layers are needed.

$$f \quad \sim \quad 1 - e^{-x/\epsilon} - e^{(x-1)/\epsilon}$$

- Example 2.   Consider

$$\epsilon^2 f'' + f = 1 \quad \text{in } 0 < x < 1 \quad \text{with} \quad f = 0 \quad \text{at } x = 0 \text{ and } 1$$

While the same stretching ϵ^α with $\alpha = 1$ produces a possible equation for the inner, there are however no decaying solutions in either direction. Hence it is not possible to add any boundary layers at $x = 0$ and $x = 1$ to help the candidate for the outer $f \sim 1$ satisfy the boundary conditions. The exact solution is

$$f \quad = \quad 1 - \frac{\sin(x/\epsilon) + \sin((1-x)/\epsilon)}{\sin(1/\epsilon)}$$

which shows that the boundary scaling is applicable all the way across the domain.

- Example 3.   Consider

$$\epsilon^2 f'' + 2f(1 - f^2) = 0 \quad \text{in } -1 < x < 1$$

$$\text{with} \quad f = -1 \quad \text{at } x = -1 \quad \text{and} \quad f = 1 \quad \text{at } x = 1$$

The solution is

$$f \sim \tanh(x/\epsilon)$$

which has a thin region of high gradients in the middle of the range near $x = 0$. The example demonstrates that the boundary layers are not always near to boundaries.

Exercise 5.6.   Find two terms in ϵ in the outer region, having matched to the inner solutions at both boundaries for

$$\epsilon^2 y'''' - y'' = -1 \quad \text{in } -1 < x < 1$$

$$\text{with } y = y' = 0 \quad \text{at } x = -1 \text{ and } 1$$

Exercise 5.7. (Cole)   The function $y(x, \epsilon)$ satisfies

$$\epsilon y'' + yy' - y = 0 \quad \text{in } 0 \le x \le 1$$

and is subject to the boundary condition $y = 0$ at $x = 0$ and $y = 3$ at $x = 1$. Assuming that there is a boundary layer only near $x = 0$, find the leading order terms in the outer and inner approximations and match them.

Exercise 5.8. (Cole)   Reconsider the equation of exercise 5.7 but now apply boundary conditions $y = -\frac{3}{4}$ at $x = 0$ and $y = \frac{5}{4}$ at $x = 1$. The boundary layer has moved to an intermediate position which is determined by the property of the inner that y jumps within the boundary layer from $-M$ to M, for some value M. Find the leading order matched asymptotic expansions.

5.1.8   Composite approximation

Now the outer expansion breaks down as $x \to 0$ because it lacks the $e^{-\xi}$ terms, while the inner expansion breaks down because it expresses $h^{(n)}(\epsilon\xi)$ as a power series in ϵ. Thus by correcting either of these two faults we can construct a uniformly valid asymptotic approximation

$$f(x, \epsilon) \sim 1 - e^{-x/\epsilon} +$$

$$\sum_0^M (-\epsilon)^n \left\{ h^{(n)}(x) - h^{(n)}(1) - \left[h^{(n)}(0) - h^{(n)}(1) \right] e^{-x/\epsilon} \right\}$$

This is called a composite approximation. Because it is not of the Poincaré form, it is not unique.

In general composite approximations can be formed by adding the inner and the outer and then subtracting their common form in the overlap. So with Van Dyke's matching rule we can form a composite limiting operator

$$C_{P,Q}f \;=\; E_P f + H_Q f - E_P H_Q f$$

Note that, if either the outer or the inner operators are applied to the composite, then we recover the outer or inner respectively:

$$E_P C_{P,Q}f = E_P f \qquad \text{and} \qquad H_Q C_{P,Q}f = H_Q f$$

When using Van Dyke's rule to do the matching, it takes little extra effort to form the composite. Thus using §5.1.5 we have for $P = Q = 0$

$$
\begin{aligned}
C_{0,0}f \;&=\; [h(x) - h(1) + 1] + [h(0) - h(1) + 1]\left(1 - e^{-\xi}\right)\\
&\quad - [h(0) - h(1) + 1]\\
&=\; h(x) - h(1) + 1 - [h(0) - h(1) + 1]\,e^{-x/\epsilon}
\end{aligned}
$$

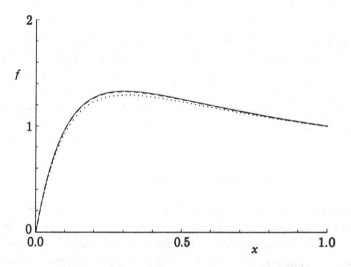

Fig. 5.3   The composite approximations for $h(x) = e^{-x}$ and $\epsilon = 0.1$. The continuous curve is the exact solution. The leading order composite approximation $C_{0,0}f$ is given by the dotted curve. The dashed curve, which is virtually superimposed on the exact solution, gives the higher composite approximation $C_{1,1}f$.

Composite approximations are often quantitatively better, i.e. with numerical values for x and ϵ, than either the inner or the outer. Composite approximations are also useful when proving the asymptoticness, because they should satisfy the governing equation and boundary conditions to ord$[\epsilon^{\min(P,Q)}]$. Some variations of the methods of matched asymptotic expansions seek from the outset a composite approximation. As some way is needed to avoid the non-uniqueness of the non-Poincaré form, these variant methods are restricted to special classes of problem.

Exercise 5.9. Find the composite $C_{1,1}f$.

Exercise 5.10. Show that $E_P C_{P,Q} f = E_P f$ and $H_Q C_{P,Q} f = H_Q f$.

Exercise 5.11. (Van Dyke)   Calculate three terms of the outer solution of

$$(1+\epsilon)x^2 y' = \epsilon \left((1-\epsilon)xy^2 - (1+\epsilon)x + y^3 + 2\epsilon y^2 \right) \quad \text{in } 0 < x < 1$$

with $y(1) = 1$. Locate the non-uniformity of the asymptoticness, and hence the rescaling for an inner region. Thence find two terms for this inner solution.

Exercise 5.12. (Van Dyke)   Consider the following problem which has an outer, an inner and an inner-inner inside the inner (called a triple deck problem)

$$x^3 y' = \epsilon \left((1+\epsilon)x + 2\epsilon^2 \right) y^2 \quad \text{in } 0 < x < 1$$

with $y(1) = 1 - \epsilon$. Calculate two terms of the outer, then two of the inner, and finally one for the inner-inner. At each stage find the rescaling required for the next layer by examining the non-uniformity of the asymptoticness in the current layer.

5.2   Logarithms

We now progress to a more advanced topic in matched asymptotic expansions. This involves logarithms. While we will find that there are two regions,   each with its own asymptotic approximation which must be matched together, the governing equation no longer gives an immediate hint of the existence of the two regions, because the small parameter does not multiply the highest derivative.

5.2.1   The problem and initial observations

We consider a model problem which looks like a problem for heat conduction outside a sphere with a small nonlinear heat source. With ϵ small and positive, let $f(r,\epsilon)$ be governed by

$$f_{rr} + \tfrac{2}{r} f_r + \epsilon f f_r = 0 \qquad \text{in } r > 1$$
$$\text{with } f = 0 \text{ at } r = 1$$
$$\text{and } f \to 1 \text{ as } r \to \infty$$

First we try naively a regular perturbation expansion for f with r fixed as $\epsilon \searrow 0$. We therefore pose formally

$$f(r,\epsilon) \sim f_0(r) + \epsilon f_2(r)$$

Note that f_1 has been omitted, because I know that this problem is not straightforward. Substituting into the governing equation and comparing coefficients of ϵ^n yields a sequence of problems.

At ϵ^0: $f_0'' + \tfrac{2}{r} f_0' = 0$ with $f_0(0) = 0$ and $f_0 \to 1$ as $r \to \infty$ with solution

$$f_0 = 1 - \frac{1}{r}$$

At ϵ^1: $f_2'' + \tfrac{2}{r} f_2' = -f_0 f_0'$ with $f_2(0) = 0$ and $f_2 \to 0$ as $r \to \infty$
The governing equation for f_2 can be rearranged to

$$\frac{1}{r^2}\left(r^2 f_2'\right)' = -\frac{1}{r^2} + \frac{1}{r^3}$$

with a solution satisfying the boundary condition at $r = 1$,

$$f_2 = -\ln r - \frac{\ln r}{r} + A_2\left(1 - \frac{1}{r}\right)$$

There is clearly trouble here, because the condition at infinity cannot be satisfied by any choice of the free constant A_2.

At this stage one might doubt that the problem has a solution: although the linearised problem is known to be well-posed, there is no supporting general theory which says that the nonlinear problem must have a solution. In a totally new problem one would probably retreat to a numerical solution of the problem, and only proceed with an asymptotic analysis when there is some evidence that a solution does exist.

The trouble with our naive expansion is that it is not uniformly asymptotic at large r. When r is large, we note that the main term in the equation $f_0'' \sim 2/r^3$ while the small nonlinear $\epsilon f_0 f_0' \sim \epsilon/r^2$. Thus the nonlinear term cannot be viewed as a small correction at $r = \text{ord}(\epsilon^{-1})$.

The difficulty at large r can be examined by introducing a rescaling $\rho = \epsilon r$. Further when $r = \text{ord}(\epsilon^{-1})$, our naive expansion suggests that $f = 1 + \text{ord}(\epsilon \ln \frac{1}{\epsilon}) + \text{ord}(\epsilon)$, and so we try the asymptotic sequence 1, $\epsilon \ln \frac{1}{\epsilon}$ and ϵ.

5.2.2   Approximation for r fixed as $\epsilon \searrow 0$

It is tempting to call this the outer approximation, because it is the solution for the unstretched variable. Unfortunately this region is inside the region with the stretched variable $\rho = \epsilon r$ fixed as $\epsilon \searrow 0$. The names outer and inner will therefore not be used in this problem.

We formally pose a Poincaré expansion in the unstretched variable,

$$f(r, \epsilon) \quad \sim \quad \left(1 - \frac{1}{r}\right) + \epsilon \ln \tfrac{1}{\epsilon} f_1(r) + \epsilon f_2(r)$$

where the obvious leading order term from §5.2.1 has been substituted.

The next function, f_1, satisfies the linearised equation. The solution satisfying the boundary condition at $r = 1$ is therefore

$$f_1 \quad = \quad A_1 \left(1 - \frac{1}{r}\right)$$

The constant A_1 cannot be determined by applying the condition at infinity directly, because that is outside the region of r fixed as $\epsilon \searrow 0$. Instead this unknown will be determined by matching with an expansion valid in the region ρ fixed as $\epsilon \searrow 0$ to which the condition at infinity can be applied.

The second correction f_2 is the same as that found in §5.2.1 with the constant A_2 to be determined now by the matching.

5.2.3   Approximation for $\rho = \epsilon r$ fixed as $\epsilon \searrow 0$

With this stretched variable the governing equation becomes

$$f_{\rho\rho} + \tfrac{2}{\rho} f_\rho + f f_\rho = 0$$

This strictly nonlinear equation is tractable only because f is very near to 1 at large r. Thus we can formally pose a Poincaré expansion

$$f(r, \epsilon) \quad \sim \quad 1 + \epsilon \ln \tfrac{1}{\epsilon} g_1(\rho) + \epsilon g_2(\rho)$$

Both g_1 and g_2 satisfy the same equation

$$g'' + \left(\frac{2}{\rho} + 1\right) g' = 0$$

i.e.

$$(\rho^2 e^\rho g')' = 0$$

Thus applying the condition that $g \to 0$ as $\rho \to \infty$, we find

$$g_i(\rho) \;=\; B_i \int_\rho^\infty \frac{e^{-\tau}}{\tau^2}\, d\tau$$

with constants B_1 and B_2 to be determined by matching. The integral above can be expressed in terms of the exponential integral; it is $E_2(\rho)/\rho$.

In preparation for matching, we need to know the behaviour of the integral in $g_i(\rho)$ for small ρ:

$$\int_\rho^\infty \frac{e^{-\tau}}{\tau^2}\, d\tau \;\sim\; \frac{1}{\rho} + (\ln\rho + \gamma - 1) - \tfrac{1}{2}\rho + o(\rho) \quad \text{as } \rho \to 0$$

in which γ is the Euler constant 0.57722.

5.2.4   Matching by intermediate variable

Introducing the intermediate variable $\eta = \epsilon^\alpha r = \rho/\epsilon^{1-\alpha}$ with $0 < \alpha < 1$, we re-express the r-approximation and the ρ-approximation in terms of η and then take the intermediate limit of η fixed as $\epsilon \searrow 0$.

$$
\begin{aligned}
r\text{-approximation} \;=\;& \left(1 - \frac{\epsilon^\alpha}{\eta}\right) \\
& + \epsilon \ln \tfrac{1}{\epsilon} A_1 \left(1 - \frac{\epsilon^\alpha}{\eta}\right) \\
& + \epsilon \left[-\alpha \ln \tfrac{1}{\epsilon} - \ln\eta + A_2 - \alpha \ln \tfrac{1}{\epsilon} \frac{\epsilon^\alpha}{\eta} - \epsilon^\alpha \frac{\ln\eta + A_2}{\eta} \right] + \cdots
\end{aligned}
$$

$$
\begin{aligned}
\rho\text{-approximation} \;=\;& 1 \\
& + \epsilon \ln \tfrac{1}{\epsilon} B_1 \left[\frac{\epsilon^{\alpha-1}}{\eta} + (\alpha - 1) \ln \tfrac{1}{\epsilon} + \ln\eta + \gamma - 1 + \cdots \right] \\
& + \epsilon B_2 \left[\frac{\epsilon^{\alpha-1}}{\eta} + (\alpha - 1) \ln \tfrac{1}{\epsilon} + \ln\eta + \gamma - 1 + \cdots \right] + \cdots
\end{aligned}
$$

These two expressions have the same form, and by forcing them to be identical we determine the constants of integration. Matching at sequential orders we find at ϵ^0: $\quad 1 = 1, \quad$ i.e. we started the r-approximation correctly

at $\epsilon^\alpha \ln \frac{1}{\epsilon}$: $\quad 0 = B_1 \frac{1}{\eta}, \qquad\qquad\qquad\qquad$ i.e. $B_1 = 0$

at ϵ^α: $\quad -\frac{1}{\eta} = B_2 \frac{1}{\eta}, \qquad\qquad\qquad$ i.e. $B_2 = -1$

at $\epsilon \ln \frac{1}{\epsilon}$: $\quad A_1 - \alpha = B_2(\alpha - 1), \qquad$ i.e. $A_1 = 1$

at ϵ: $\quad -\ln \eta + A_2 = B_2(\ln \eta + \gamma - 1), \quad$ i.e. $A_2 = 1 - \gamma$

Note in the last but one equation that once the constants A_1 and B_2 are determined for one value of α then the equation becomes true for all α, i.e. true for all intermediate limits.

We have now determined the solution. For r fixed

$$ f \sim \left(1 - \frac{1}{r}\right) + \epsilon \ln \frac{1}{\epsilon} \left(1 - \frac{1}{r}\right) + \epsilon \left[-\ln r - \frac{\ln r}{r} + (1 - \gamma)\left(1 - \frac{1}{r}\right)\right] $$

while for ρ fixed

$$ f \sim 1 + 0\epsilon \ln \frac{1}{\epsilon} - \epsilon \int_\rho^\infty \frac{e^{-\tau}}{\tau^2} \, d\tau $$

5.2.5 Further terms

To understand the correction terms to the above solution, it is necessary to review where the present terms have come from. The leading order, ord(1), term in the r-region forces through the small nonlinear term in the governing equation the correction ord(ϵ). This term contained an unmatchable $\ln r$, which called for the introduction of the ρ-region. In the ρ-region, the forced ord(ϵ) term behaved like $\ln \rho$ as $\rho \to 0$. Matching then required the introduction of an ord($\epsilon \ln \frac{1}{\epsilon}$) term in the r-region. Note that this unexpected ord($\epsilon \ln \frac{1}{\epsilon}$) term in the r-region is not directly forced by the field equation there – it is a homogeneous or eigensolution of the linearised equation. Such a term which is forced by the matching is sometimes called a *switchback*. These logarithmic terms naturally occur in particular integrals of differential equations,

$$ \int_{\text{ord}(1)}^{\text{ord}(\frac{1}{\epsilon})} \frac{dr}{r} $$

Now turning to the correction terms. The $\epsilon \ln \frac{1}{\epsilon}$ term in the r-region will force through the small nonlinear term in the governing equation a correction $\epsilon^2 \ln \frac{1}{\epsilon}$ which must have a $\ln r$ behaviour at large r, just like the ord(1) term of which it is a copy. The process of matching will then require an ord($\epsilon^2 \ln \frac{1}{\epsilon}$) term in the ρ-region with a $\ln \rho$ behaviour at small ρ and thence an $\epsilon^2[\ln\frac{1}{\epsilon}]^2$ term in the r-region. Thus we can expect corrections $\mathrm{ord}(\epsilon^2[\ln\frac{1}{\epsilon}]^2)$, $\mathrm{ord}(\epsilon^2\ln\frac{1}{\epsilon})$ and $\mathrm{ord}(\epsilon^2)$.

5.2.6   Failure of Van Dyke's matching rule

If we take the E operator for the r-limit and the H operator for the ρ-limit, then Van Dyke's matching rule works at $P = Q = 0$ and at $P = Q = 2$. It fails, however, at $P = Q = 1$, i.e. when retaining the terms $\mathrm{ord}(1)$ and $\mathrm{ord}(\epsilon\ln\frac{1}{\epsilon})$.

$$
\begin{aligned}
H_1 E_1 f &= H_1\left\{\left(1-\frac{1}{r}\right)+\epsilon\ln\frac{1}{\epsilon}\left(1-\frac{1}{r}\right)\right\}\\
&= H_1\left\{\left(1-\frac{\epsilon}{\rho}\right)+\epsilon\ln\frac{1}{\epsilon}\left(1-\frac{\epsilon}{\rho}\right)\right\}\\
&= 1+\epsilon\ln\frac{1}{\epsilon}\\
E_1 H_1 f &= E_1\left[1+\epsilon\ln\frac{1}{\epsilon}0\right]\\
&= 1
\end{aligned}
$$

The trouble is that the $\epsilon\ln r$ term changes its order. Consider a $\ln r$ term on its own. Let

$$
\varphi(r,\epsilon)\ \equiv\ 0+\frac{\ln r}{\ln\frac{1}{\epsilon}}\ \equiv\ 1+\frac{\ln\rho}{\ln\frac{1}{\epsilon}}
$$

Then Van Dyke's matching rule with the asymptotic sequence 1, $[\ln\frac{1}{\epsilon}]^{-1}$ will work for this function for $(P,Q) = (0,1)$, $(1,0)$ and $(1,1)$, and it fails for $(0,0)$. A more general term $(\ln r)^n$ will lead to *some* failures near to the diagonal where $|P-Q| < n$. A term like $1/\ln r$ or $\ln\ln r$ leads to more serious trouble.

When applying Van Dyke's rule, good advice is to match only at a break where the power of ϵ changes, if that is possible. Thus in our problem with the asymptotic sequence

$$
1;\quad \epsilon\ln\tfrac{1}{\epsilon},\ \epsilon;\quad \epsilon^2[\ln\tfrac{1}{\epsilon}]^2,\ \epsilon^2\ln\tfrac{1}{\epsilon},\ \epsilon^2;\quad \ldots
$$

it is wisest to match with P and Q corresponding to the semicolons. Moreover, because $\ln\frac{1}{\epsilon}$ is rarely large for typical small values of ϵ, it is always necessary to calculate all the terms which only differ by a factor of $\ln\frac{1}{\epsilon}$.

5.2.7   Composite approximation

Because Van Dyke's rule fails for $P = Q = 1$, there is no composite $C_{1,1}f$. But the other composites exist.

$$C_{0,0}f = 1 - \frac{1}{r}$$

$$C_{2,2}f = 1 - \epsilon\left[\int_{\epsilon r}^{\infty} \frac{e^{-\tau}}{\tau^2}\, d\tau + \frac{1}{r}\left(\ln\tfrac{1}{\epsilon} + 1 - \gamma + \ln r\right)\right]$$

Exercise 5.13.   The function $f(r, \epsilon)$ satisfies the equation

$$f_{rr} + \tfrac{2}{r}f_r + \tfrac{1}{2}\epsilon^2(1 - f^2) = 0 \qquad \text{in } r > 1$$

and is subject to the boundary conditions

$$f = 0 \quad \text{at} \quad r = 1 \qquad \text{and} \qquad f \to 1 \quad \text{as} \quad r \to \infty$$

Using the asymptotic sequence 1, ϵ, $\epsilon^2 \ln\frac{1}{\epsilon}$, ϵ^2, obtain asymptotic expansions for f at fixed r as $\epsilon \searrow 0$ and at fixed $\rho = \epsilon r$ as $\epsilon \searrow 0$. Match the expansions using the intermediate variable $\eta = \epsilon^\alpha r$ with $0 < \alpha < 1$.

You may quote that the solution to the equation

$$y_{xx} + \tfrac{2}{x}y_x - y = \frac{e^{-2x}}{x^2}$$

subject to the condition $y \to 0$ as $x \to \infty$ is

$$y = A\frac{e^{-x}}{x} + \frac{1}{2x}\int_x^\infty \frac{e^{-x-t} - e^{x-3t}}{t}\, dt$$

with A a constant. Further as $x \to 0$

$$y \sim \frac{2A + \ln 3}{2x} + \ln x - A + \gamma + \tfrac{1}{2}\ln 3 - 1$$

Exercise 5.14.   The function $f(r, \epsilon)$ satisfies the equation

$$f_{rr} + \tfrac{3}{2r}f_r + \epsilon f f_r = 0 \qquad \text{in } r > 1$$

and is subject to the boundary conditions

$$f = 0 \quad \text{at} \quad r = 1 \qquad \text{and} \qquad f \to 1 \quad \text{as} \quad r \to \infty$$

and with $\epsilon > 0$. Obtain an asymptotic expansion for f at fixed r as $\epsilon \to 0$ in the asymptotic sequence 1, $\epsilon^{1/2}$, $\epsilon \ln\frac{1}{\epsilon}$, ϵ; and an asymptotic expansion for f at fixed $\rho = \epsilon r$ as $\epsilon \to 0$ in the sequence 1, $\epsilon^{1/2}$, ϵ. Match these expansions.

The general solution $y(x)$ with $y \to 0$ as $x \to \infty$ of

$$\left(x^{3/2}e^x y'\right)' \;=\; E_{3/2}(x) \;=\; \int_x^\infty t^{-3/2}e^{-t}\,dt$$

is $y(x) = BE_{3/2} + F(x)$ with B an arbitrary constant, and that as $x \to 0$,

$$y \;=\; B\left(2x^{-1/2} - 2\sqrt{\pi} + 2x^{1/2} + O(x^{3/2})\right) + \left(4\ln x + C + O(x^{1/2})\right)$$

where C is a numerical constant.

5.2.8   A worse problem

This second model problem is like heat conduction outside a cylinder with a small nonlinear heat source. With ϵ small and positive, let $f(r, \epsilon)$ be governed by

$$f_{rr} + \tfrac{1}{r}f_r + \epsilon f f_r \;=\; 0 \qquad \text{in } r > 1$$
$$\text{with } f = 0 \text{ at } r = 1$$
$$\text{and } f \to 1 \text{ as } r \to \infty$$

First we try a regular perturbation expansion for r fixed as $\epsilon \searrow 0$, i.e. we pose formally

$$f(r, \epsilon) \;\sim\; f_0(r) + \epsilon f_1(r)$$

At $\mathrm{ord}(\epsilon^0)$ we find that

$$f_0(r) \;=\; A_0 \ln r$$

satisfying the boundary condition at $r = 1$. But the condition at infinity cannot be satisfied with any choice of the free constant A_0. If we were to continue to $\mathrm{ord}(\epsilon^1)$, we would have a worse problem satisfying the condition at infinity with

$$f_1(r) \;=\; -A_0^2 \left(r \ln r - 2r + 2\right) + A_1 \ln r$$

At this stage one might wonder whether the problem for f is well-posed and a solution exists. The linearised problem is certainly ill-posed. It happens that the nonlinear problem does have a solution; see figure 5.4.

First we note that the rescaling $\rho = \epsilon r$ is applicable to the new equation because it only differs from the previous equation in a numerical factor. The troublesome condition at infinity is then in the region ρ fixed as $\epsilon \searrow 0$. Now as f must be something like the above f_0 in

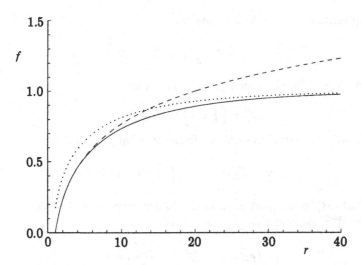

Fig. 5.4   The solution of the problem of §5.2.8 with $\epsilon = 0.05$. The continuous curve gives the exact solution obtained numerically. The dashed curve is the leading order approximation of the r-fixed region, while the dotted curve gives the approximation for the ρ-fixed region of the leading order unity plus one correction term.

the r-region and it must also be ord(1) in the ρ-region, we see that A_0 should be ord($[\ln \frac{1}{\epsilon}]^{-1}$). This suggests an asymptotic sequence 1, $[\ln \frac{1}{\epsilon}]^{-1}$, $[\ln \frac{1}{\epsilon}]^{-2}, \ldots$ with no leading order term in the r-region.

Approximation for r fixed as $\epsilon \searrow 0$

We now start again with the new, less obvious asymptotic sequence. Thus we pose formally a Poincaré expansion

$$f(r, \epsilon) \quad \sim \quad \frac{1}{\ln \frac{1}{\epsilon}} f_1(r) \; + \; \frac{1}{[\ln \frac{1}{\epsilon}]^2} f_2(r)$$

Then each $f_i(r)$ satisfies the same linearised equation, with solution satisfying the boundary condition at $r = 1$

$$f_i(r) \; = \; A_i \ln r$$

Approximation for ρ fixed as $\epsilon \searrow 0$

In terms of the stretched variable $\rho = \epsilon r$ the governing equation becomes

$$f_{\rho\rho} + \tfrac{1}{\rho} f_\rho + f f_\rho \; = \; 0$$

Posing formally a Poincaré expansion

$$f(r, \epsilon) \quad \sim \quad 1 + \frac{1}{\ln \frac{1}{\epsilon}} g_1(\rho) + \frac{1}{[\ln \frac{1}{\epsilon}]^2} g_2(\rho)$$

we find at ord($[\ln \frac{1}{\epsilon}]^{-1}$) that g_1 is governed by

$$g_1'' + \left(\frac{1}{\rho} + 1\right) g_1' \quad = \quad 0$$

with a solution satisfying the condition at infinity $g_1 \to 0$ as $\rho \to \infty$,

$$g_1 \quad = \quad B_1 \int_\rho^\infty \frac{e^{-\tau}}{\tau} \, d\tau \quad = \quad B_1 E_1(\rho)$$

in which B_1 is a constant and E_1 is the exponential integral. At ord($[\ln \frac{1}{\epsilon}]^{-2}$) we find that g_2 is governed by

$$(\rho e^\rho g_2')' = B_1^2 \int_\rho^\infty \frac{e^{-\tau}}{\tau} \, d\tau$$

with solution satisfying the condition $g_2 \to 0$ as $\rho \to \infty$,

$$g_2 \quad = \quad B_2 E_1(\rho) + B_1^2 \left(2E_1(2\rho) - e^{-\rho} E_1(\rho)\right)$$

In preparation for matching, we need the behaviour of the g_i as $\rho \to 0$. Now

$$E_1(\rho) \quad \sim \quad -\ln \rho - \gamma + \rho$$
$$2E_1(2\rho) - e^{-\rho} E_1(\rho) \quad \sim \quad -\ln \rho - \gamma - \ln 4 - \rho \ln \rho + (3 - \gamma)\rho$$

Matching by intermediate variable

With $\eta = \epsilon^\alpha r = \rho/\epsilon^{1-\alpha}$ with $0 < \alpha < 1$, our r-approximation becomes

$$0 + \frac{1}{\ln \frac{1}{\epsilon}} A_1 \left(\alpha \ln \frac{1}{\epsilon} + \ln \eta\right) + \frac{1}{[\ln \frac{1}{\epsilon}]^2} A_2 \left(\alpha \ln \frac{1}{\epsilon} + \ln \eta\right) + \cdots$$

while our ρ-approximation becomes

$$1 + \frac{1}{\ln \frac{1}{\epsilon}} B_1 \left[-(\alpha - 1) \ln \frac{1}{\epsilon} - \ln \eta - \gamma + \cdots\right]$$

$$+ \frac{1}{[\ln \frac{1}{\epsilon}]^2} B_2 \left[-(\alpha - 1) \ln \frac{1}{\epsilon} - \ln \eta - \gamma + \cdots\right]$$

$$+ \frac{1}{[\ln \frac{1}{\epsilon}]^2} B_1^2 \left[-(\alpha - 1) \ln \frac{1}{\epsilon} - \ln \eta - \gamma - \ln 4 + \cdots\right] + \cdots$$

Comparing terms of sequential order

at $\ln \frac{1}{\epsilon}$: $\alpha A_1 = 1 - B_1(\alpha - 1)$

This is true for all α, i.e. for all intermediate limits which means the entire overlap region, if

$$B_1 = -1 \quad \text{and} \quad A_1 = 1$$

At $[\ln \frac{1}{\epsilon}]^{-1}$: $A_1 \ln \eta + \alpha A_2 = -B_1 \ln \eta - B_1 \gamma - (\alpha - 1)B_2 - (\alpha - 1)B_1^2$

Substituting the known A_1 and B_1, and again requiring the matching to work for all intermediate limits gives

$$A_2 = \gamma \quad \text{and} \quad B_2 = -1 - \gamma$$

Thus all the unknowns have been determined without proceeding to ord$([\ln \frac{1}{\epsilon}]^{-2})$.

Matching by Van Dyke's rule

This fails when $P = Q$, as explained in §5.2.5. Thus if we take E for the r-limit and H for the ρ-limit, we find

$E_0 H_0 f = 1$ and $H_0 E_0 f = 0$,   which is impossible

$E_0 H_1 f = 1 + B_1$ and $H_1 E_0 f = 0$,   i.e. $B_1 = -1$ correctly

$E_1 H_0 f = 1$ and $H_0 E_1 f = A_1$,   i.e. $A_1 = 1$ correctly

$E_1 H_1 f = 1 + B_1 + [\ln \frac{1}{\epsilon}]^{-1} B_1(-\ln r - \gamma)$

and $H_1 E_1 f = A_1 + [\ln \frac{1}{\epsilon}]^{-1} A_1 \ln \rho = A_1 - 1 + [\ln \frac{1}{\epsilon}]^{-1} \ln r$

In this last case if we put $A_1 = 1$ and $B_1 = -1$ then the ord(1) terms match and the $\ln r$ dependence of the ord$([\ln \frac{1}{\epsilon}]^{-1})$ matches,   but the constant term at this order does not match.

Exercise 5.15.   Check that Van Dyke's rule works for $(P, Q) = (0, 2)$, $(1, 2)$, $(2, 1)$ and $(2, 0)$, but fails for $(2, 2)$.

5.2.9   A terrible problem

This third model equation is unusually difficult. The function is something like $\ln(1 + \ln r / \ln \frac{1}{\epsilon})$. For this function Van Dyke's rule fails for all values of P and Q. The intermediate variable method of matching also struggles, because the leading order part of an infinite number of terms must be calculated before the matching can be made successfully. It is unusual to find such a difficult problem in practice.

With ϵ small and positive, let $f(r, \epsilon)$ be governed by

$$f_{rr} + \tfrac{1}{r}f_r + f_r^2 + \epsilon f f_r = 0 \qquad \text{in } r > 1$$

$$\text{with } f = 0 \text{ at } r = 1$$

$$\text{and } f \to 1 \text{ as } r \to \infty$$

The new extra nonlinear term appears to be an ord(1) disruption to the equation studied in §5.2.8. In the r-region, however, f was small, ord$([\ln\tfrac{1}{\epsilon}]^{-1})$, and so this quadratic term would be smaller than the first two linear terms. And in the ρ-region f was very nearly 1 with a deviation, and hence gradient, of order $[\ln\tfrac{1}{\epsilon}]^{-1}$. Thus we can anticipate that f has the same asymptotic scalings in this new problem despite the new term.

Approximation for r fixed as $\epsilon \searrow 0$

We start by posing a Poincaré expansion in inverse powers of $\ln\tfrac{1}{\epsilon}$ which starts with the first power

$$f(r, \epsilon) \quad \sim \quad \frac{1}{\ln\tfrac{1}{\epsilon}} f_1(r) \; + \; \frac{1}{[\ln\tfrac{1}{\epsilon}]^2} f_2(r) \; + \; \frac{1}{[\ln\tfrac{1}{\epsilon}]^3} f_3(r)$$

Substituting into the governing equation and the boundary condition at $r = 1$, and comparing coefficients of $[\ln\tfrac{1}{\epsilon}]^{-n}$, we find

at $[\ln\tfrac{1}{\epsilon}]^{-1}$: $f_1'' + \tfrac{1}{r}f_1' = 0$ with $f_1 = 0$ at $r = 1$

with solution $f_1 = A_1 \ln r$

at $[\ln\tfrac{1}{\epsilon}]^{-2}$: $f_2'' + \tfrac{1}{r}f_2' = -f_1'^2 = \tfrac{1}{r}(rf_2')' = -A_1^2 \tfrac{1}{r^2}$

with solution $f_2 = A_2 \ln r - \tfrac{1}{2}A_1^2 \ln^2 r$

at $[\ln\tfrac{1}{\epsilon}]^{-3}$: $f_3'' + \tfrac{1}{r}f_3' = -2f_1'f_2' = -A_1\tfrac{2}{r}\left(-A_1^2\tfrac{\ln r}{r} + A_2\tfrac{1}{r}\right)$

with solution $f_3 = A_3 \ln r + \tfrac{1}{3}A_1^3 \ln^3 r - A_1 A_2 \ln^2 r$.

From the structure of the above problems we can see that the general term f_n will have leading order behaviour as $r \to \infty$

$$f_n \quad \sim \quad (-)^n \left(-\tfrac{1}{n}A_1^n \ln^n r + A_1^{n-2} A_2 \ln^{n-1} r\right)$$

This can be checked with an induction argument. Note that these leading order parts of the f_n can be summed to

$$\ln\left[1 + \left(\frac{A_1}{\ln\tfrac{1}{\epsilon}} + \frac{A_2}{[\ln\tfrac{1}{\epsilon}]^2} + \cdots\right)\ln r\right]$$

which satisfies

$$f_{rr} + \tfrac{1}{r}f_r + f_r^2 = 0$$

Approximation for ρ fixed as $\epsilon \searrow 0$

In terms of the ρ variable the governing equation becomes

$$f_{\rho\rho} + \tfrac{1}{\rho}f_\rho + f_\rho^2 + ff_\rho = 0$$

We formally pose a Poincaré expansion

$$f(r,\epsilon) \quad \sim \quad 1 + \frac{1}{\ln\frac{1}{\epsilon}}g_1(\rho) + \frac{1}{[\ln\frac{1}{\epsilon}]^2}g_2(\rho) + \frac{1}{[\ln\frac{1}{\epsilon}]^3}g_3(\rho)$$

Substituting into the above stretched form of the governing equation, and comparing coefficients of $[\ln\frac{1}{\epsilon}]^{-n}$, we find

at $[\ln\frac{1}{\epsilon}]^{-1}$: $\quad g_1'' + \tfrac{1}{\rho}g_1' + g_1' = \tfrac{1}{\rho}e^{-\rho}(\rho e^\rho g_1')' = 0$

Integrating and imposing the condition that $g_1 \to 0$ as $\rho \to \infty$ yields

$$g_1 \quad = \quad B_1 \int_\rho^\infty \frac{e^{-\tau}}{\tau}\,d\tau \quad = \quad B_1 E_1(\rho)$$

At $[\ln\frac{1}{\epsilon}]^{-2}$: $\quad g_2'' + \tfrac{1}{\rho}g_2' + g_2' = -g_1'^2 - g_1 g_1'$, $\quad$ i.e.

$$(\rho e^\rho g_2')' \quad = \quad -B_1^2 \left(\frac{e^{-\rho}}{\rho} - \int_\rho^\infty \frac{e^{-\tau}}{\tau}\,d\tau \right)$$

with solution satisfying the condition at infinity

$$g_2 \quad = \quad B_2 E_1(\rho) + B_1^2 \left(2E_1(2\rho) - \tfrac{1}{2}E_1^2(\rho) - e^{-\rho}E_1(\rho) \right)$$

In preparation for matching, we need the behaviour as $\rho \to 0$,

$$g_1 \sim B_1(-\ln\rho - \gamma)$$
$$g_2 \sim B_2(-\ln\rho - \gamma) + B_1^2\left(-\tfrac{1}{2}\ln^2\rho - (\gamma+1)\ln\rho - \tfrac{1}{2}\gamma^2 - \gamma - \ln 4\right)$$

We note that the leading order behaviour in g_2 comes from

$$g_2'' + \tfrac{1}{\rho}g_2' \sim -g_1'^2 \sim -\frac{B_1^2}{\rho^2}$$

Hence the leading order behaviour in the following g_3 will come from

$$g_3'' + \tfrac{1}{\rho}g_3' \sim -2g_1'g_2' \sim -2B_1^3\frac{\ln\rho}{\rho^2} - \left(2B_1^3(\gamma+1) + 2B_1B_2\right)\frac{1}{\rho^2}$$

and so

$$g_3 \sim -\tfrac{1}{3}B_1^3\ln^3\rho - \left(B_1^3(\gamma+1) + B_1B_2\right)\ln^2\rho$$

Again it can be shown that the leading order behaviour as $\rho \to 0$ of the general term $g_n(\rho)$ is

$$g_n(\rho) \sim -\tfrac{1}{n} B_1^n \ln^n \rho - \left(B_1^n (\gamma + 1) + B_1^{n-2} B_2 \right) \ln^{n-1} \rho$$

Matching by intermediate variable

With $\eta = \epsilon^\alpha r = \rho/\epsilon^{1-\alpha}$ fixed as $\epsilon \searrow 0$, we have

$$r\text{-approximation} \quad \sim \quad \frac{1}{\ln \frac{1}{\epsilon}} A_1 \left[\alpha \ln \tfrac{1}{\epsilon} + \ln \eta \right]$$

$$+ \; \frac{1}{[\ln \frac{1}{\epsilon}]^2} \left\{ -\tfrac{1}{2} A_1^2 \left[\alpha \ln \tfrac{1}{\epsilon} + \ln \eta \right]^2 + A_2 \left[\alpha \ln \tfrac{1}{\epsilon} + \ln \eta \right] \right\}$$

$$+ \; \frac{1}{[\ln \frac{1}{\epsilon}]^3} \left\{ \tfrac{1}{3} A_1^3 \left[\alpha \ln \tfrac{1}{\epsilon} + \ln \eta \right]^3 - A_1 A_2 \left[\alpha \ln \tfrac{1}{\epsilon} + \ln \eta \right]^2 \right.$$

$$\left. + A_3 \left[\alpha \ln \tfrac{1}{\epsilon} + \ln \eta \right] \right\}$$

$$+ \; \frac{1}{[\ln \frac{1}{\epsilon}]^4} \left\{ -\tfrac{1}{4} A_1^4 \left[\alpha \ln \tfrac{1}{\epsilon} + \ln \eta \right]^4 + A_1^2 A_2 \left[\alpha \ln \tfrac{1}{\epsilon} + \ln \eta \right]^3 + \cdots \right\}$$

$$+ \; \cdots$$

and

$$\rho\text{-approximation} \sim 1 + \frac{1}{\ln \frac{1}{\epsilon}} B_1 \left[-(\alpha - 1) \ln \tfrac{1}{\epsilon} - \ln \eta - \gamma + \cdots \right]$$

$$+ \; \frac{1}{[\ln \frac{1}{\epsilon}]^2} \left\{ -B_1^2 \left(\tfrac{1}{2} \left[(\alpha - 1) \ln \tfrac{1}{\epsilon} + \ln \eta \right]^2 - (\gamma + 1) \left[(\alpha - 1) \ln \tfrac{1}{\epsilon} + \ln \eta \right] \right.\right.$$

$$\left.\left. + \tfrac{1}{2} \gamma^2 + \gamma + \ln 4 + \cdots \right) + B_2 \left[-(\alpha - 1) \ln \tfrac{1}{\epsilon} - \ln \eta - \gamma + \cdots \right] \right\}$$

$$+ \; \frac{1}{[\ln \frac{1}{\epsilon}]^3} \left\{ -B_1^3 \tfrac{1}{3} \left[(\alpha - 1) \ln \tfrac{1}{\epsilon} + \ln \eta \right]^3 - B_1^3 (\gamma + 1) \times \right.$$

$$\left. \left[(\alpha - 1) \ln \tfrac{1}{\epsilon} + \ln \eta \right]^2 - B_1 B_2 \left[(\alpha - 1) \ln \tfrac{1}{\epsilon} + \ln \eta \right]^2 + \cdots \right\}$$

$$+ \; \cdots$$

Comparing coefficients of $[\ln \frac{1}{\epsilon}]^0$, we find

$$A_1 \alpha - \tfrac{1}{2} A_1^2 \alpha^2 + \tfrac{1}{3} A_1^3 \alpha^3 - \tfrac{1}{4} A_1^4 \alpha^4 + \cdots \quad =$$

$$1 - B_1(\alpha - 1) - \tfrac{1}{2} B_1^2 (\alpha - 1)^2 - \tfrac{1}{3} B_1^3 (\alpha - 1)^3 + \cdots$$

The two sides of the equation are well known infinite series which converge for $0 < \alpha < 1$, yielding

$$\ln(1 + A_1\alpha) = 1 + \ln[1 - B_1(\alpha - 1)] = \ln(e[1 - B_1(\alpha - 1)])$$

Requiring the matching to work throughout the overlap region, i.e. for all intermediate limits, i.e. all values of α, gives

$$A_1 = e - 1 \quad \text{and} \quad B_1 = -(e-1)/e$$

Comparing coefficients of $[\ln\frac{1}{\epsilon}]^{-1}$ in the r- and ρ-approximations in the overlap region, we find

$$
\begin{aligned}
& \ln\eta\left(A_1 - \tfrac{1}{2}A_1^2 2\alpha + \tfrac{1}{3}A_1^3 3\alpha^2 - \tfrac{1}{4}A_1^4 4\alpha^3 + \cdots\right) \\
& \quad + \left(A_2\alpha - A_1 A_2\alpha^2 + A_1^2 A_2\alpha^3 + \cdots\right) \\
= \;& \ln\eta\left(-B_1 - \tfrac{1}{2}B_1^2 2(\alpha - 1) - \tfrac{1}{3}B_1^3 3(\alpha - 1)^2 + \cdots\right) \\
& \quad + \left(-B_1\gamma - B_1^2(\gamma + 1)(\alpha - 1) - B_1^3(\gamma + 1)(\alpha - 1)^2 + \cdots\right) \\
& \quad + \left(-B_2(\alpha - 1) - B_1 B_2(\alpha - 1)^2 + \cdots\right)
\end{aligned}
$$

Again summing, this is

$$
\begin{aligned}
\frac{A_1}{1 + A_1\alpha}\ln\eta + \frac{A_2\alpha}{1 + A_1\alpha} =\;& \frac{-B_1}{1 - B_1(\alpha - 1)}\ln\eta \\
& -\frac{B_1\gamma}{1 - B_1(\alpha - 1)} + \frac{B_1^2(\alpha - 1)}{1 - B_1(\alpha - 1)} - \frac{B_2(\alpha - 1)}{1 - B_1(\alpha - 1)}
\end{aligned}
$$

The $\ln\eta$ terms balance with the previously determined A_1 and B_1. The constant terms give

$$\frac{A_2}{A_1}\alpha = \gamma - B_1(\alpha - 1) + \frac{B_2}{B_1}(\alpha - 1)$$

Requiring this to be true for all α, we find

$$A_2 = \gamma(e - 1) \quad \text{and} \quad B_2 = (e-1)(e - 1 - \gamma e)/e^2$$

Note that in the matching an infinite number of terms jumped their order. It was therefore necessary to have obtained the leading order behaviour of the general terms f_n and g_n.

5.3 Slow viscous flow

The model problems of §5.2 contain the mathematically difficulty in finding the small inertial corrections to the viscous flow past a sphere and past a cylinder. These two problems are known as the Stokes–Whitehead paradoxes, and their resolution was influential in the development of the method of matched asymptotic expansions.

5.3.1   Past a sphere

Axisymmetric flow past a sphere can be described by a Stokes stream-function ψ which satisfies the Navier–Stokes equation in the form

$$\frac{\epsilon}{r^2 \sin\theta} \left(\frac{\partial\psi}{\partial\theta}\frac{\partial}{\partial r} - \frac{\partial\psi}{\partial r}\frac{\partial}{\partial\theta} + 2\cot\theta\frac{\partial\psi}{\partial r} - \frac{2}{r}\frac{\partial\psi}{\partial\theta} \right) D^2\psi$$

$$= D^2 D^2\psi \qquad \text{in } r \geq 1$$

with $\quad \psi = \dfrac{\partial\psi}{\partial r} = 0 \quad$ on $r = 1$

and $\quad \psi \to \frac{1}{2}r^2 \sin^2\theta \quad$ as $r \to \infty$

in which $\quad D^2 = \dfrac{\partial^2}{\partial r^2} + \dfrac{\sin\theta}{r^2}\dfrac{\partial}{\partial\theta}\left(\dfrac{1}{\sin\theta}\dfrac{\partial}{\partial\theta}\right)$

Approximation for r fixed as $\epsilon \searrow 0$

We formally pose an expansion

$$\psi(r,\theta;\epsilon) \quad \sim \quad \psi_0(r,\theta) + \epsilon\psi_1(r,\theta)$$

The lowest term is governed by the equation

$$D^2 D^2\psi_0 = 0$$

and is forced by the condition at infinity. Looking for a solution proportional to $\sin^2\theta$, we find possible radial dependencies r^4, r^2, r and r^{-1}. Satisfying the boundary conditions on $r = 1$,

$$\psi_0 = \frac{1}{4}\left(2r^2 - 3r + \frac{1}{r}\right)\sin^2\theta$$

Substituting this into the left hand side of the governing equation produces

$$-\epsilon\frac{9}{4}\left(\frac{2}{r^2} - \frac{3}{r^3} + \frac{1}{r^5}\right)\sin^2\theta\cos\theta$$

to force ψ_1. Again we look for a solution proportional to $\sin^2\theta\cos\theta$. To the particular integral (which is made to satisfy the boundary condition on $r = 1$) we must add a homogeneous solution, which turns out just to be a multiple of ψ_0. (Other homogeneous solutions can be added, but during the matching they will be found to have zero coefficients.)

$$\psi_1 = -\tfrac{3}{32}\left(2r^2 - 3r + 1 - \frac{1}{r} + \frac{1}{r^2}\right)\sin^2\theta\cos\theta$$

$$+ A_1\left(2r^2 - 3r + \frac{1}{r}\right)\sin^2\theta$$

But no choice of the free constant A_1 enables the condition at infinity to be satisfied at all θ.

Approximation for $\rho = \epsilon r$ fixed as $\epsilon \searrow 0$

Now in the far field at large r the above r-approximation has

$$\psi = \tfrac{1}{2}r^2\sin^2\theta + \mathrm{ord}(r, \epsilon r^2)$$

corresponding to a uniform flow plus the disturbance from a point force together with some inertial corrections. This suggests an expansion in the ρ-region

$$\psi(r, \theta; \epsilon) \sim \frac{1}{\epsilon^2}\tfrac{1}{2}\rho^2\sin^2\theta + \frac{1}{\epsilon}\Psi_1(\rho, \theta)$$

The equation governing Ψ_1 is the Oseen equation in the form

$$\left(D_\rho^2 - \cos\theta\frac{\partial}{\partial\rho} + \frac{\sin\theta}{\rho}\frac{\partial}{\partial\theta}\right)D_\rho^2\Psi_1 = 0$$

where D^2 has been modified with ρ replacing r. The homogeneous solution we need turns out to be that corresponding to a point force

$$\Psi_1 = B_1(1 + \cos\theta)\left(1 - e^{-\frac{1}{2}\rho(1-\cos\theta)}\right)$$

which can most easily be obtained in Cartesian co-ordinates by Fourier transforming the linear Oseen equation. This derivation also gives an immediate connection between B_1 and the drag. In Ψ_1 the $(1 + \cos\theta)$ factor corresponds to a source flow, while the exponential describes a wake concentrated in a region $\rho\theta^2 = \mathrm{ord}(1)$; the source being needed to remove a mass defect in the wake.

Matching by intermediate variable

With $\eta = \epsilon^\alpha r = \rho/\epsilon^{1-\alpha}$ fixed as $\epsilon \searrow 0$

$$r\text{-approximation} = \tfrac{1}{4}\left(2\epsilon^{-2\alpha}\eta^2 - 3\epsilon^{-\alpha}\eta + \cdots\right)\sin^2\theta$$

$$- \epsilon\tfrac{3}{32}\left(2\epsilon^{-2\alpha}\eta^2 + \cdots\right)\sin^2\theta\cos\theta + \epsilon A_1\left(2\epsilon^{-2\alpha}\eta^2 + \cdots\right)\sin^2\theta$$

$$+ \cdots$$

and

$$\rho\text{-approximation} \quad = \quad \frac{1}{\epsilon^2}\tfrac{1}{2}\epsilon^{2-2\alpha}\eta^2\sin^2\theta$$

$$+ \; \frac{1}{\epsilon}B_1\left(1+\cos\theta\right)\left(\tfrac{1}{2}\epsilon^{1-\alpha}\eta(1-\cos\theta) - \tfrac{1}{8}\epsilon^{2-2\alpha}\eta^2(1-\cos\theta)^2 + \cdots\right)$$

$$+ \; \cdots$$

Comparing terms of sequential order we find

at $\epsilon^{-2\alpha}$: $\tfrac{1}{2}\eta^2\sin^2\theta = \tfrac{1}{2}\eta^2\sin^2\theta$ – just uniform flow

at $\epsilon^{-\alpha}$: $-\tfrac{3}{4}\eta\sin^2\theta = \tfrac{1}{2}B_1\eta(1-\cos^2\theta)$, i.e. $B_1 = -\tfrac{3}{2}$

at $\epsilon^{1-2\alpha}$: $-\tfrac{3}{16}\eta^2\sin^2\theta\cos\theta + 2A_1\eta^2\sin^2\theta$

$$= -\tfrac{1}{8}B_1\eta^2\sin^2\theta(1-\cos\theta), \quad \text{i.e. } A_1 = \tfrac{3}{32}$$

The fact that the value of A_1 is $\tfrac{3}{8}$ times the $\tfrac{1}{4}$ in ψ_0 leads to an enhancement in the drag on the sphere by a factor $(1+\tfrac{3}{8}\epsilon)$. In effect at this order the r-region sees a uniform flow $(1+\tfrac{3}{8}\epsilon)$. The next order terms are ord $\left(\epsilon^2\ln\tfrac{1}{\epsilon}\right)$ and ord(ϵ^2).

5.3.2   Past a cylinder

For this two-dimensional flow we use a streamfunction ψ which satisfies

$$\epsilon\left(\frac{1}{r}\frac{\partial\psi}{\partial\theta}\frac{\partial}{\partial r} - \frac{1}{r}\frac{\partial\psi}{\partial r}\frac{\partial}{\partial\theta}\right)\nabla^2\psi \quad = \quad \nabla^2\nabla^2\psi \qquad \text{in } r \geq 1$$

with $\psi = \dfrac{\partial\psi}{\partial r} = 0$ on $r = 1$

and $\psi \to r\sin\theta$ as $r \to \infty$

Approximation for r fixed as $\epsilon \searrow 0$

Now the lowest order term must be governed by $\nabla^2\nabla^2\psi_0 = 0$. Looking for a solution proportional to the forcing $\sin\theta$, we find possible radial dependencies r^3, $r\ln r$, r and r^{-1}. It is not possible to satisfy the conditions on $r = 1$ and $r \to \infty$: the least unpleasant solution at infinity satisfying the boundary conditions at $r = 1$ is

$$f_*(r,\theta) \quad = \quad \left(r\ln r - \tfrac{1}{2}r + \frac{1}{2r}\right)\sin\theta$$

As in §5.2.8, we expect this solution not to apply when $\rho = \epsilon r = \text{ord}(1)$, and so we need to multiply the above f_* by $[\ln\tfrac{1}{\epsilon}]^{-1}$ to reduce the magnitude correctly. This leads to an expansion in powers of $[\ln\tfrac{1}{\epsilon}]^{-1}$. All the terms in the r-region must then be proportional to the above f_*.

$$\psi(r,\theta;\epsilon) \quad \sim \quad \frac{1}{\ln\frac{1}{\epsilon}}A_1 f_*(r,\theta) + \frac{1}{[\ln\frac{1}{\epsilon}]^2}A_2 f_*(r,\theta)$$

Approximation for $\rho = \epsilon r$ fixed as $\epsilon \searrow 0$

In this region the flow is the uniform flow plus a small ord($[\ln\frac{1}{\epsilon}]^{-1}$) correction,

$$\psi(r,\theta;\epsilon) \quad \sim \quad \frac{1}{\epsilon}\rho\sin\theta + \frac{1}{\epsilon\ln\frac{1}{\epsilon}}\Psi_1(\rho,\theta)$$

The equation governing Ψ_1 is Oseen's

$$\left(\nabla^2_\rho - \cos\theta\frac{\partial}{\partial\rho} + \frac{\sin\theta}{\rho}\frac{\partial}{\partial\theta}\right)\nabla^2_\rho\Psi_1 \;=\; 0$$

Again we need the homogeneous solution corresponding to a point force which can be obtained by Fourier transforming (but see also the hint for the exercise at the end of the section). The inversion for Ψ_1 cannot be expressed in closed form, although the vorticity $\nabla^2\Psi_1$ and the velocity $\nabla\Psi_1$ can be. From the Fourier transform, one can extract the behaviour of Ψ_1 as $\rho \to 0$

$$\Psi_1 \quad \sim \quad B_1\rho(\ln\rho - \ln 4 + \gamma - 1)\sin\theta$$

Matching by intermediate variable

With $\eta = \epsilon^\alpha r = \rho/\epsilon^{1-\alpha}$ fixed as $\epsilon \searrow 0$

$$r\text{-approximation} \quad = \quad 0 + \frac{1}{\ln\frac{1}{\epsilon}}A_1\epsilon^{-\alpha}\eta\sin\theta\left(\alpha\ln\frac{1}{\epsilon} + \ln\eta - \frac{1}{2} + \cdots\right)$$

$$+ \frac{1}{[\ln\frac{1}{\epsilon}]^2}A_2\epsilon^{-\alpha}\eta\sin\theta\left(\alpha\ln\frac{1}{\epsilon} + \ln\eta - \frac{1}{2} + \cdots\right)$$

$$+ \cdots$$

and

$$\rho\text{-approximation} \quad = \quad \frac{1}{\epsilon}\epsilon^{1-\alpha}\eta\sin\theta$$

$$+ \frac{1}{\epsilon\ln\frac{1}{\epsilon}}B_1\left(\epsilon^{1-\alpha}\eta\sin\theta\left[(\alpha-1)\ln\frac{1}{\epsilon} + \ln\eta - \ln 4 + \gamma - 1\right] + \cdots\right)$$

$$+ \cdots$$

Comparing terms of sequential orders,

at $\epsilon^{-\alpha}$: $\qquad \alpha A_1\eta\sin\theta = \eta\sin\theta + B_1(\alpha-1)\eta\sin\theta$

which is true for all intermediate limits if

$$A_1 = B_1 = 1$$

At $\epsilon^{-\alpha}[\ln\frac{1}{\epsilon}]^{-1}$: $A_1\left(\ln\eta - \frac{1}{2}\right)\eta\sin\theta + \alpha A_2\eta\sin\theta$

$$= B_1(\ln\eta - \ln 4 + \gamma - 1)\eta\sin\theta + B_2(\alpha - 1)\eta\sin\theta$$

in which B_2 is the coefficient of a similar homogeneous solution at ord($\epsilon^{-1}[\ln\frac{1}{\epsilon}]^{-2}$) in the ρ-region, which happens to dominate the particular integral. Again matching for arbitrary intermediate limit, we find

$$A_2 = B_2 = -\ln 4 + \gamma - \frac{1}{2}$$

Note that because $\ln\frac{1}{\epsilon}$ is rarely large numerically, some people combine $[\ln\frac{1}{\epsilon}]^{-1}A_1$ and $[\ln\frac{1}{\epsilon}]^{-2}A_2$ when they occur in the drag to form $1/[\ln(4/\epsilon) - \gamma + \frac{1}{2}]$. At higher orders there are alternative methods of improving the convergence of a series – see chapter 8.

Exercise 5.16. Consider the heat transfer from a cylinder in a weak potential flow. Thus solve for $T(\mathbf{x}, \epsilon)$ which satisfies

$$\epsilon\mathbf{u}\cdot\nabla T = \nabla^2 T \quad \text{in } r \geq 1$$

with $T = 1$ on $r = 1$

and $T \to 0$ as $r \to \infty$

where $\mathbf{u} = \mathbf{U}\left(1 + \frac{1}{r^2}\right) - \mathbf{x}\frac{2\,(\mathbf{U}\cdot\mathbf{x})}{r^4}$

Then calculate

$$\int_{r=1}\frac{\partial T}{\partial n}\,dA$$

Hint: The substitution $\varphi = \psi e^{x/2}$ turns the Oseen equation governing φ

$$\left(\frac{\partial}{\partial x} - \nabla^2\right)\varphi$$

into the simpler equation

$$\left(\tfrac{1}{4} - \nabla^2\right)\psi$$

Exercise 5.17. Now try the case of weak potential flow past a sphere, with

$$\mathbf{u} = \mathbf{U}\left(1 + \frac{1}{2r^3}\right) - \mathbf{x}\frac{3\,(\mathbf{U}\cdot\mathbf{x})}{2r^5}$$

5.4   Slender body theory

The method of matched asymptotic expansions is used in problems which have two (or more) naturally occurring length scales; asymptotic expansions being made for each of the scales, the expansions then being matched in order to determine some constants of integration. In the problems considered so far, the governing equation generated the second length scale; that of the thin boundary layer in §5.1 and that of the far field in §§5.2 and 5.3. It is also possible for the basic geometry to have more than one natural length scale. Long slender bodies considered in this section have the scales of their width and their length. Thus on the smaller scale the bodies appear to be nearly infinitely long, quasi-uniform, finite diameter cylinders, while on the longer scale they appear to have a finite length, but to be vanishingly thin. Other problems whose geometry has more than one natural length scale include the interaction between greatly separated particles (with the scales of their size and their separation – see exercise 5.18 at the end of this section) and waves scattering off small scale inhomogeneities (with the scales of the inhomogeneities and the wavelength).

For simplicity we will only study slender bodies with straight centre-lines, although in §5.4.1 we will have non-circular cross-sections. Let ϵ be the (small) slenderness; then in cylindrical polar co-ordinates the surface of the body may be taken as

$$r = \epsilon R(\theta, z) \qquad \text{in } |z| \le 1$$

5.4.1   Electrical capacitance

We must solve for the potential $\varphi(r, \theta, z; \epsilon)$ which satisfies

$$\nabla^2 \varphi = 0 \qquad \text{outside the body}$$

and is subject to boundary conditions

$$\varphi = 1 \text{ on the body and } \varphi \to 0 \text{ at infinity}$$

The capacitance can then be evaluated as

$$- \int \frac{\partial \varphi}{\partial n} \, dA$$

Approximation for $\rho = r/\epsilon$ fixed as $\epsilon \searrow 0$. In this scaling the governing equation becomes

$$\frac{1}{\rho}\frac{\partial}{\partial\rho}\left(\rho\frac{\partial\varphi}{\partial\rho}\right) + \frac{1}{\rho^2}\frac{\partial^2\varphi}{\partial\theta^2} + \epsilon^2\frac{\partial^2\varphi}{\partial z^2} = 0 \quad \text{in } \rho \geq R(\theta,z)$$

This suggests an expansion in ϵ^2. Here we look at just the leading order term in such an expansion; a correction term will be calculated for the problem in the following section. For the leading order term, the variable z does not occur in the field equation. We therefore have to solve a two-dimensional potential problem, i.e. at this order and in this scaling the cylinder appears to be infinitely long. The shape of the cylinder is specified by $R(\theta,z)$ in which we must view z just as a parameter. Applying the boundary condition on the surface, we have a general solution

$$\varphi_0(r,\theta,z) = 1 + Q(z)\left(\frac{1}{2\pi}\ln\frac{\rho}{R_*(z)} + \sum_{n=1}^{\infty} B_n(z)\rho^{-n}\frac{\cos}{\sin}n\theta\right)$$

where $R_*(z)$ and the $B_n(z)$ depend on the local shape $R(\theta,z)$ and where $Q(z)$ has to be found from matching. The matching effectively applies the condition at infinity which cannot be applied in the strictly two-dimensional potential problem.

The quantity R_* is known in two-dimensional potential theory as the effective or equivalent radius of the cross-section.

- For a circular cross-section $R(\theta,z) = R(z)$,

$$R_*(z) = R(z) \quad \text{and} \quad B_n = 0.$$

- For an elliptical cross-section with semi-diameters $a(z)$ and $b(z)$,

$$R_*(z) = \tfrac{1}{2}(a+b).$$

This can be derived using the conformal map

$$\zeta = \tfrac{1}{2}\rho e^{i\theta} + \tfrac{1}{2}\sqrt{\rho^2 e^{2i\theta} - a^2 + b^2}$$

which takes the ellipse into the circle $|\zeta| = \tfrac{1}{2}(a+b)$. Hence the complex potential is

$$1 + \frac{1}{2\pi}Q\ln\frac{2\zeta}{a+b}$$

But as $\rho \to \infty$, $\zeta \sim \rho e^{i\theta}\{1 - \tfrac{1}{4}(a^2-b^2)e^{-2i\theta}\rho^{-2}\}$. Substituting this into the potential yields $R_* = \tfrac{1}{2}(a+b)$ and $B_1 = 0$ and $B_2 = \tfrac{1}{4}(a^2-b^2)$.

- For a cross-section which is a star composed of n points of zero width and length a,

$$R_*(z) = a(z)2^{-2/n}$$

This can also be derived using a conformal map. This time use $\zeta = \rho^n e^{in\theta}$ to map the star into a degenerate ellipse with $b = 0$.

Approximation for r fixed as $\epsilon \searrow 0$. In this scaling the body has a finite length, but nearly no thickness. We thus see the body as a line of distributed charges $Q(z)$, with weaker higher order poles $B_n(z)$. Thus we try as a first approximation in this r-region

$$\varphi(r,\theta,z;\epsilon) \quad \sim \quad -\int_{-1}^{1} \frac{Q(z';\epsilon)\,dz'}{4\pi\sqrt{r^2 + (z-z')^2}}$$

In preparation for matching, our r-approximation takes the form

$$\varphi(r,\theta,z;\epsilon) \quad = \quad -\frac{1}{4\pi}Q(z)2\ln\frac{1}{r} + O(Q) \qquad \text{as } r \to 0$$

Matching with an intermediate variable $\eta = r/\epsilon^\alpha = \epsilon^{1-\alpha}\rho$ fixed as $\epsilon \searrow 0$.

r-approximation $\quad = \quad \frac{1}{2\pi}Q(z)\left(-\alpha\ln\frac{1}{\epsilon} + \ln\eta + O(1)\right) + \cdots$

ρ-approximation $\quad = \quad 1 + \frac{1}{2\pi}Q(z)\Big((1-\alpha)\ln\frac{1}{\epsilon}$

$$+ \ln\frac{\eta}{R_*} + \epsilon^{1-\alpha}B_1\eta^{-1}\cos\theta + \cdots \Big) + \cdots$$

Thus at leading order $Q(z) = -2\pi[\ln\frac{1}{\epsilon}]^{-1}$ with an error of order $[\ln\frac{1}{\epsilon}]^{-2}$. If we expand Q in a series of inverse powers of $\ln\frac{1}{\epsilon}$

$$Q(z;\epsilon) \quad \sim \quad -\frac{1}{\ln\frac{1}{\epsilon}}2\pi + \frac{1}{[\ln\frac{1}{\epsilon}]^2}Q_2(z)$$

then from the earlier solution of the integral equation in §3.5 we have in the r-region

$$\varphi \sim \frac{1}{\ln\frac{1}{\epsilon}}\frac{1}{2}\ln\frac{4(1-z^2)}{r^2} - \frac{1}{[\ln\frac{1}{\epsilon}]^2}\frac{1}{4\pi}Q_2\left(2\ln\frac{1}{r} + O(1)\right) \qquad \text{as } r \to 0$$

So matching again

r-approximation $\quad = \quad -\frac{1}{\ln\frac{1}{\epsilon}}\left(-\alpha\ln\frac{1}{\epsilon} + \ln\eta - \frac{1}{2}\ln 4(1-z^2)\right)$

$$+ \frac{1}{[\ln\frac{1}{\epsilon}]^2}\frac{1}{2\pi}Q_2\left(-\alpha\ln\frac{1}{\epsilon} + \cdots\right) + \cdots$$

ρ-approximation $\quad = \quad 1$

$$+ \; \frac{1}{2\pi} \left(-\frac{1}{[\ln \frac{1}{\epsilon}]^2} \pi + \frac{1}{[\ln \frac{1}{\epsilon}]^2} Q_2(z) \right) \left((1 - \alpha) \ln \tfrac{1}{\epsilon} + \ln \frac{\eta}{R_*} + \cdots \right) + \cdots$$

we find that at leading order $\alpha = 1 - 1 + \alpha$, and at $[\ln \frac{1}{\epsilon}]^{-1}$ matching is successful if

$$Q_2(z) \quad = \quad 2\pi \ln \frac{2\sqrt{1 - z^2}}{R_*(z)}$$

The electrical capacitance of the slender body can now be evaluated:

$$- \int \frac{\partial \varphi}{\partial n} \, dA \quad = \quad - \int_{-1}^{1} Q(z) \, dz$$

$$\sim \quad \frac{4\pi}{\ln \frac{1}{\epsilon}} - \frac{2\pi}{[\ln \frac{1}{\epsilon}]^2} \int_{-1}^{1} \ln \frac{2\sqrt{1 - z^2}}{R_*(z)} \, dz$$

This expression shows that the capacitance depends only weakly on the shape, involving just the logarithm of the slenderness at the leading order, and in the correction just an integral of the cross-section.

We note that prior to the matching the asymptotic theory had error terms ord(ϵ^2). During the matching it was necessary to introduce an expansion in powers of $[\ln \frac{1}{\epsilon}]^{-1}$, which is rarely small in practice. Now the matching is equivalent to solving the integral equation

$$1 \quad = \quad - \int_{-1}^{1} \frac{Q(z') \, dz'}{\sqrt{\epsilon^2 R_*(z)^2 + (z - z')^2}}$$

i.e. in the r-region the body appears to be a circular cylinder of radius R_*. To avoid the expansion in logarithms, this integral equation can be solved numerically (for particular values of ϵ).

For slender bodies with centre lines which are not straight, the form of the solution in the ρ-region will not be changed by the curvature until ord(ϵ^2). In the r-region the body will be represented by charges $Q(z)$ distributed along the curved centre line. In the evaluation of this r-region integral for matching, the leading order term of the $\ln \frac{1}{\epsilon}$ expansion is found to be unaffected by the curvature, but the correction, $[\ln \frac{1}{\epsilon}]^{-1}$ smaller, does depend on the global shape of the centre line. Thus Q_2 would be different.

5.4.2   Axisymmetric potential flow

For simplicity we now restrict attention to a circular cross-section. Irrotational incompressible flow can be described by a velocity potential

$\varphi(r, z; \epsilon)$ which satisfies

$$\nabla^2 \varphi = 0 \qquad \text{outside the body}$$

$$\text{with} \quad \frac{\partial \varphi}{\partial n} = 0 \qquad \text{on the surface of the body}$$

The condition at infinity corresponding to a uniform flow in the axial direction is

$$\varphi \to z$$

Before we start, we need to note that the normal (not a unit normal) to the surface $r = \epsilon R(z)$ is $(1, -\epsilon R')$, where the prime denotes differentiation with respect to z. Thus the boundary condition on the surface of the body becomes

$$\frac{\partial \varphi}{\partial r} - \epsilon R' \frac{\partial \varphi}{\partial z} = 0$$

Approximation for $\rho = r/\epsilon$ fixed as $\epsilon \searrow 0$. In this scaling the field equation and boundary condition become

$$\frac{1}{\rho} \frac{\partial}{\partial \rho} \left(\rho \frac{\partial \varphi}{\partial \rho} \right) + \epsilon^2 \frac{\partial^2 \varphi}{\partial z^2} = 0 \qquad \text{in } \rho \geq R(z)$$

$$\text{and} \quad \frac{\partial \varphi}{\partial \rho} - \epsilon^2 R' \frac{\partial \varphi}{\partial z} = 0 \qquad \text{on } \rho = R(z)$$

These suggest an expansion in ϵ^2, the first term corresponding to an infinite cylinder, while the second term includes some effect of the tapering ($R' \neq 0$). In the leading order term, there is just a constant (which may depend on the parameter z), because the boundary condition rules out a $\ln \rho$. The ϵ^2 correction term is forced by the field equation and the boundary condition. Thus

$$\varphi(r, z; \epsilon) \quad \sim \quad A_0(z) + \epsilon^2 \left(-\tfrac{1}{4} A_0'' \rho^2 + A_2(z) + B_2(z) \ln \rho \right)$$

where

$$B_2 = R \left(A_0' R' + \tfrac{1}{2} A_0'' R \right)$$

Approximation for r fixed as $\epsilon \searrow 0$. In this scaling the body appears as a line distribution of sources $2\pi \epsilon^2 B_2(z)$, i.e.

$$\varphi(r, z) \quad \sim \quad z - \int_{-1}^{1} \frac{\epsilon^2 B_2(z') \, dz'}{2\sqrt{r^2 + (z - z')^2}}$$

Preparing for matching, this r-approximation takes the form

$$\varphi \quad \sim \quad z - \epsilon^2 B_2(z) \left(\ln \frac{1}{r} + O(1) \right) \qquad \text{as } r \to 0$$

Matching with an intermediate variable $\eta = r/\epsilon^a = \rho\epsilon^{1-a}$ fixed as $\epsilon \searrow 0$.

$$r\text{-approximation} \quad = \quad z + B_2(z)\left(-\alpha \ln \tfrac{1}{\epsilon} + \ln \eta + O(1)\right) + \cdots$$

$$\rho\text{-approximation} \quad = \quad A_0(z) + \epsilon^2 \left(\tfrac{1}{4} - A_0'' \epsilon^{2\alpha - 2}\eta^2 + A_2(z)\right.$$

$$\left. + B_2(z)\left(-(\alpha - 1)\ln \tfrac{1}{\epsilon} + \ln \eta\right)\right) + \cdots$$

Matching at ord(1) we find

$$A_0 = z \quad \text{and so} \quad B_2 = RR'$$

The ord($\epsilon^{2\alpha}$) term therefore evaporates. At ord($\epsilon^2 \ln \tfrac{1}{\epsilon}$) we find A_2 must be slightly larger than expected, with $A_2 = B_2 \ln \tfrac{1}{\epsilon}$ plus $O(1)$ corrections. However B_2 remains unchanged up to ord($\epsilon^2 \ln \tfrac{1}{\epsilon}$).

The value $RR' = (\tfrac{1}{2}R^2)'$ for B_2 can be interpreted as follows. The volume flux of the undisturbed flow across the cross-sectional area of the body is $\pi\epsilon^2 R^2$. The internally distributed sources $2\pi\epsilon^2 B_2$ ensure that there is no flow into the body. These sources are needed where the cross-sectional area changes.

Exercise 5.18. Another problem where the geometry presents two length scales is that of the electrical capacitance of a pair of thin parallel wires. Let the wires have radii ϵ and be centred at $x = \pm\tfrac{1}{2}$, i.e. have surfaces

$$(x \pm \tfrac{1}{2})^2 + y^2 \; = \; \epsilon^2$$

Then outside both these surfaces one needs to solve $\nabla^2 \varphi = 0$ for the potential φ subject to the boundary conditions that φ is a constant, $V(\epsilon)$, independent of position around the boundary on one of the wires, and $-V$ on the other. If the charge on the wires is taken to be ∓ 1 per unit length, then the capacitance is $2V$.

Now at leading order the electrical charges are uniformly distributed around the surface of each wire, i.e.

$$\varphi \quad \sim \quad -\frac{1}{2\pi}\ln\sqrt{(x - \tfrac{1}{2})^2 + y^2} + \frac{1}{2\pi}\ln\sqrt{(x + \tfrac{1}{2})^2 + y^2}$$

Examine the potential in the neighbourhood of one of the wires by expanding the above potential to ord(ϵ^2) using the stretched variables $x = \tfrac{1}{2} + \epsilon\xi$ and $y = \epsilon\eta$. Now construct an improved approximation for the potential by adding dipoles (like $(x \pm \tfrac{1}{2})/[(x \pm \tfrac{1}{2})^2 + y^2]$) and quadrupoles (like $[y^2 - (x \pm \tfrac{1}{2})^2]/[(x \pm \tfrac{1}{2})^2 + y^2]^2$) of appropriate magnitudes to ensure that the potential is constant around the surfaces to ord(ϵ^2). (The magnitude of the monopoles is the total charge which does not vary from unity.) Hence find the capacitance $2V(\epsilon)$ to ord(ϵ^4).

5.5 Moon space ship problem

In the Grand Tour space mission, a space craft passed the major planets in sequence, receiving much of its kinetic energy with respect to the Sun from being redirected as it flew past each planet. Very high accuracy was needed in predicting the path of the space craft in order that it would fly closely past the second and subsequent planets. Integrating the primitive equations of motion for the space craft in the gravitational field of the Sun and its orbiting planets does not produce sufficient accuracy on the largest computers. Lagerstrom and Kevorkian showed in 1963 that the problem could be tackled with matched asymptotic expansions based on the smallness of the mass of the planets compared with that of the Sun, and that the desired accuracy could then be achieved in the calculations using a small computer. The idea is that the space craft moves along a classical elliptical or hyperbolic orbit around the Sun until it comes near to a planet. It then orbits the planet and on leaving the planet it sets off on a new orbit around the Sun. The matching involves extracting from the old orbit the energy and impact parameter for the new orbit.

Here we study very briefly a model problem in which the space craft moves in the gravitational field of the Earth at $(x, y) = (0, 0)$ and an ϵ-mass Moon which is fixed at $(1, 0)$, starting the space craft at $t = 0$ from the Earth with its escape velocity (i.e. potential plus kinetic energy is zero) in nearly the direction of the Moon, dy/dx along the path is ϵk with k a constant. The governing equations are

$$\ddot{x} = -\frac{x}{(x^2 + y^2)^{3/2}} - \epsilon \frac{x - 1}{((x - 1)^2 + y^2)^{3/2}}$$

$$\ddot{y} = -\frac{y}{(x^2 + y^2)^{3/2}} - \epsilon \frac{y}{((x - 1)^2 + y^2)^{3/2}}$$

Approximation for orbit about the Earth. In the 'outer' approximation we pose an expansion

$$x(t, \epsilon) \sim x_0(t) + \epsilon x_1(t) \qquad \text{and} \qquad y(t, \epsilon) \sim 0 + \epsilon y_1(t)$$

in which ϵx_1 is the first effect of the Moon, and the ϵy_1 term does not in fact feel the Moon but is small because there is little motion in the

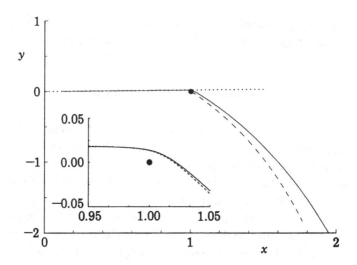

Fig. 5.5   The trajectory of the space craft for $\epsilon = 0.02$ and $k = 1$. The continuous curve is the exact solution obtained numerically. The approximate orbits around the Earth are given by the dotted and the dashed curves. The inset gives the fine details in the neighbourhood of the Moon at $x = 1$ and $y = 0$, with the dashed curve being the approximate orbit around the Moon.

y-direction initially. For the leading order approximation, we have

$$\ddot{x}_0 = -\frac{1}{x_0^2}$$

with the first integral

$$\tfrac{1}{2}\dot{x}_0^2 - \frac{1}{x_0} = 0$$

choosing the constant of integration to be zero by the initial condition that the space craft sets out with precisely its escape velocity. Integrating again

$$x_0 = \left(\tfrac{9}{2}\right)^{1/3} t^{2/3}$$

As claimed above, y_1 is not affected by the Moon since it is governed by

$$\ddot{y}_1 = -\frac{y_1}{x_0^3}$$

with simple solution

$$y_1 = k x_0(t)$$

i.e. the space craft keeps going in the same direction as its initial direction (conservation of angular momentum). The equation for x_1 is

$$\ddot{x}_1 \;=\; \frac{2x_1}{x_0^3} + \frac{1}{(1-x_0)^2}$$

As $t \nearrow t_* = \sqrt{2}/3$, $x_0 \nearrow 1$ and the above equation gives

$$x_1 \;\sim\; -\tfrac{1}{2}\ln(t_* - t)$$

and so the orbit about the Earth breaks down as the craft approaches the Moon. A complete solution for x_1 can be obtained most easily by recombining x_0 and ϵx_1 into $\tilde{x}$, their recombined equation being twice integrable to

$$\tfrac{2}{3}\tilde{x}^{3/2} - \epsilon\left(\tfrac{1}{2}\ln\frac{1+\tilde{x}^{1/2}}{1-\tilde{x}^{1/2}} - \tilde{x}^{1/2} - \tfrac{1}{3}\tilde{x}^{3/2}\right) \;=\; \sqrt{2}t$$

and so

$$x_1 \;=\; \tfrac{1}{2}x_0^{-1/2}\ln\frac{1+x_0^{1/2}}{1-x_0^{1/2}} - 1 - \tfrac{1}{3}x_0$$

Approximation for orbit around the Moon. The scaling for this 'inner' region is found from the requirements that the velocity $\Delta x/\Delta t = \mathrm{ord}(1)$ in order to match to the above outer and that the acceleration is due to the ϵ mass $\Delta x/(\Delta t)^2 = \mathrm{ord}(\epsilon/(\Delta x)^2)$. Thus $\Delta x = \Delta y = \Delta t = \epsilon$, so we rescale

$$x = 1 + \epsilon\xi, \qquad y = \epsilon\eta \qquad \text{and} \qquad t = t_* + \epsilon\tau$$

producing governing equations

$$\xi_{\tau\tau} \;=\; -\frac{\xi}{(\xi^2+\eta^2)^{3/2}} - \epsilon + O(\epsilon^2)$$

$$\eta_{\tau\tau} \;=\; -\frac{\eta}{(\xi^2+\eta^2)^{3/2}} + O(\epsilon^2)$$

i.e. the Moon's attraction dominates and the Earth gives rise to a relatively small (the ϵ term) uniform gravitation acceleration in the neighbourhood of the Moon. At lowest order, we have a classical central orbit problem with solution

$$(\xi^2 + \eta^2)^{1/2} \;=\; \frac{b\sin\alpha}{\cos\alpha + \cos(\alpha - \tan^{-1}(\eta/\xi))}$$

with impact parameter b and deflection $\pi - 2\alpha$, which are related to the velocity at infinity v_* by

$$v_*^2 = \frac{\tan \alpha}{b}$$

At higher orders we would find that the constant acceleration from the Earth leads to a switch-back $\epsilon \ln \frac{1}{\epsilon}$ term.

Matching. As we approach the Moon, the orbit around the Earth has

$$\dot{x}_0 \sim \sqrt{2} \quad \text{and} \quad y_1 \sim k \quad \text{as } t \nearrow t_*$$

while as we start the orbit about the Moon

$$\xi_\tau \sim v_* \quad \text{and} \quad \eta \sim b \quad \text{as } \tau \searrow -\infty$$

Hence $b = k$ and $v_* = \sqrt{2}$. The deflection angle for the orbit around the Moon is then $\pi - 2\tan^{-1}(2k)$. At higher orders there is a time delay $\mathrm{ord}(\epsilon \ln \frac{1}{\epsilon})$.

Second orbit around the Earth. This starts from near $x = 1$ and $y = 0$ with a velocity at the end of the orbit around the Moon v_* in the direction 2α. Thus we need a central orbit around the Earth through $x = 1$ and $y = 0$ with energy $\frac{1}{2}v_*^2 - 1 = 0$ and angular momentum $\sqrt{2}\sin 2\alpha = 4\sqrt{2}k/(1+4k^2)$. Note that the energy is unchanged by the encounter but the angular momentum is increased. (The energy with respect to the Earth would have been increased if we had considered a moving Moon.) The desired second orbit about the Earth is then

$$(x^2 + y^2)^{1/2} = \frac{32k^2}{(1 + 4k^2)^2 \left[1 - \cos\left(\tan^{-1}\frac{y}{x} - 4\tan^{-1}(2k)\right)\right]}$$

Exercise 5.19.   Solve for $a(t, \epsilon)$ and $r(t, \epsilon)$ which are governed by

$$\dot{a} = -a\left(\frac{1}{a} - \epsilon r^2\right)$$

$$\dot{r} = r\left(\frac{1}{a} - \epsilon r^2\right) - 1$$

with $a = r = 1$ at $t = 0$.

Find the leading order approximation for when $a, r, t = \mathrm{ord}(1)$. Note that this approximation breaks down when $t = 1-\mathrm{ord}(\epsilon)$ with $a = \mathrm{ord}(\epsilon)$ and $r = \mathrm{ord}(\epsilon^{-1})$. With the rescaling $a = \mathrm{ord}(\epsilon)$, $r = \mathrm{ord}(\epsilon^{-1})$ and $\Delta t = \mathrm{ord}(\epsilon^{-1})$, find the new leading order approximation and match crudely. [The tough part of this problem is to match properly through a transition region with $a = \mathrm{ord}(\epsilon)$, $r = \mathrm{ord}(\epsilon^{-1})$ and $\Delta t = \mathrm{ord}(\epsilon)$.]

5.6   van der Pol relaxation oscillator

This problem is typical of one of the more complicated applications of
the method of matched asymptotic expansions. To make the details look
simpler the formal matching with an intermediate variable is abandoned.
Along with the less formal approach, however, there is the new idea
of examining the way one expansion breaks down in order to find the
rescaling appropriate for the next region.

The van der Pol oscillator is governed by the equation

$$\ddot{x} + \mu\dot{x}(x^2 - 1) + x = 0$$

This oscillator has large nonlinear friction which is negative in $|x| <
1$ and positive in $|x| > 1$. As a result the trivial solution $x \equiv 0$ is
unstable, while large amplitudes are damped. Thus all solutions tend
to a finite amplitude oscillation, which balances energy losses in $|x| > 1$
with energy gains in $|x| < 1$. We now try to find the form of this so-
called relaxation oscillation or limit cycle as $\mu \to \infty$. Setting $\mu = \infty$ in
the equation shows that this problem is singular. From computations
(see figure 5.6) it is found that the oscillation consists of fast phases with
$\Delta t = \text{ord}(\mu^{-1})$ in which the large friction or anti-friction is balanced by
inertia and slow phases with $\Delta t = \text{ord}(\mu)$ in which the large friction
balances the restoring force. We now briefly construct a solution for this

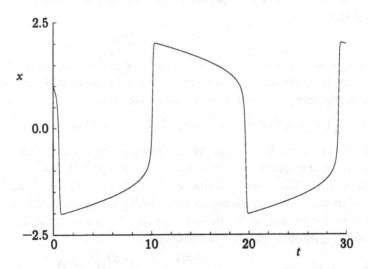

Fig. 5.6   The relaxation oscillation of the van der Pol equation with $\mu = 10$.

relaxation oscillation by matched asymptotic expansions. Without loss of generality we may start at $t = 0$ with $x = 1$, because the relaxation oscillation must pass between the damping $|x| > 1$ and the anti-damping $|x| < 1$.

Slow phase. Rescaling with $t = \mu T$, the governing equations become

$$\mu^{-2} x_{TT} + x_T (x^2 - 1) + x = 0$$

This suggests an expansion

$$x(t, \mu) \sim X_0(T) + \mu^{-2} X_1(T)$$

At μ^0: $X_0'(X_0^2 - 1) + X_0 = 0$ with $X_0 = 1$ at $T = 0$
with implicit solution

$$T = \ln X_0 - \tfrac{1}{2}(X_0^2 - 1)$$

As $T \nearrow 0$, this solution breaks down because the right hand side is negative, with

$$X_0 \sim 1 + (-T)^{1/2} \qquad \text{as } T \nearrow 0$$

At μ^{-2}: $X_1'(X_0^2 - 1) + 2X_0' X_0 X_1 + X_1 = -X_0''$
This can be solved in the form $X_1 = f(X_0)$, but here we need only the behaviour of the particular integral as $T \nearrow 0$, because it is the particular integral which breaks the asymptoticness of the expansion and calls for another balance in the equation. (This particular integral represents the effect of inertia which was neglected in the first approximation to the slow phase.)

$$X_1 \sim \tfrac{1}{4}(-T)^{-1} \qquad \text{as } T \nearrow 0$$

Matching forwards, we reconstruct our slow phase solution as it approaches its breakdown at $T = 0$ using the original time variable (rather than properly using a general intermediate time variable)

$$x \sim \left(1 + (-\mu^{-1}t)^{1/2} + \cdots\right) + \mu^{-2}\left(\tfrac{1}{4}(-\mu^{-1}t)^{-1} + \cdots\right) + \cdots$$

The correction term is as large as the leading order term minus 1 (which is the important measure) when $t = \text{ord}(\mu^{-1/3})$ where $x = 1 + \text{ord}(\mu^{-2/3})$. The trouble is that as $T \nearrow 0$, $x \searrow 1$, so the coefficient of friction $(x^2 - 1)$ drops, so the velocity increases, and so the inertia is no longer negligible. In fact with the above scaling, all three terms in the governing equation are $\text{ord}(1)$:

$$\ddot{x} \;:\; \mu \dot{x}(x^2 - 1) \;:\; x \;=\; \frac{\mu^{-2/3}}{(\mu^{-1/3})^2} \;:\; \mu \frac{\mu^{-2/3}}{\mu^{-1/3}}\left(\mu^{-2/3}\right) \;:\; 1$$

Transition phase. With the rescaling suggested above, $t = \mu^{-1/3}s$ and $x = 1 + \mu^{-2/3}z$, the governing equation becomes

$$z_{ss} + 2z_s z + 1 + \mu^{-2/3}(z_s z^2 + z) = 0$$

which suggests an expansion in a revised sequence

$$x(t,\mu) \sim 1 + \mu^{-2/3}z_1(s) + \mu^{-4/3}z_2(s)$$

Matching backwards into the slow phase, we find

$$z_1 \quad \sim \quad (-s)^{1/2} + \tfrac{1}{4}(-s)^{-1} \qquad \text{as } s \searrow -\infty$$

At $\mu^{-2/3}$: $\qquad z_1'' + 2z_1' z_1 + 1 = 0$

This can be integrated once choosing the constant by matching backwards:

$$z_1' + z_1^2 + s = 0$$

This Ricatti equation can be turned into the Airy equation with the substitution $z_1 = \zeta'/\zeta$:

$$\zeta'' + s\zeta = 0$$

The general solution can be expressed in terms of $K_{1/3}$ and $I_{1/3}$ for $s < 0$. Matching backwards rules out $I_{1/3}$, so for $s < 0$

$$\zeta = (-s)^{1/2} K_{1/3}\left(\tfrac{2}{3}(-s)^{3/2}\right)$$

This solution continues into $s > 0$ in the form

$$\tfrac{1}{\sqrt{3}}(s)^{1/2}\left[J_{1/3}\left(\tfrac{2}{3}s^{3/2}\right) + J_{-1/3}\left(\tfrac{2}{3}s^{3/2}\right)\right]$$

Because $z = \zeta'/\zeta$, there is trouble with $z \searrow -\infty$ as $s \nearrow s_0 = 2.34$, the first root of $J_{1/3} + J_{-1/3} = 0$. From the equation for ζ we find

$$z_1 \quad \sim \quad -\frac{1}{s_0 - s} + \tfrac{1}{3}s_0(s_0 - s) \qquad \text{as } s \nearrow s_0$$

Matching forward, we reconstruct the above solution in the transition phase in terms of the original time variable as it approaches its breakdown:

$$x \quad \sim \quad 1 + \mu^{-2/3}\left[-\frac{\mu^{-1/3}}{\mu^{-1/3}s_0 - t} + \mu^{1/3}\tfrac{1}{3}s_0(\mu^{-1/3}s_0 - t)\right]$$

The asymptoticness is broken when $\mu^{-1/3}s_0 - t = \text{ord}(\mu^{-1})$ where $x = \text{ord}(1)$.

Fast phase. With the scaling suggested above, $t = \mu^{-1/3}s_0 + \mu^{-1}\tau$, the governing equation becomes

$$x_{\tau\tau} + x_\tau(x^2 - 1) + \mu^{-2}x = 0$$

Matching backwards into the end of the transition phase

$$x \quad \sim \quad 1 + \tau^{-1} - \mu^{-4/3}\tau\tfrac{1}{3}s_0 \qquad \text{as } \tau \searrow -\infty$$

This suggests an asymptotic expansion

$$x(t,\mu) \quad \sim \quad x_0(\tau) + \mu^{-4/3}x_1(\tau) + \mu^{-2}x_2(\tau)$$

At μ^0: $\qquad x_0'' + x_0'(x_0^2 - 1) = 0$

Integrating the equation once and choosing the constant of integration by matching backwards to $x_0 \sim 1 + \tau^{-1}$ as $\tau \searrow -\infty$

$$x_0' + \tfrac{1}{3}x_0^3 - x_0 = -\tfrac{2}{3}$$

Integrating again and matching backwards yields an implicit solution

$$\tfrac{1}{3}\ln\frac{2 + x_0}{1 - x_0} + \frac{1}{1 - x_0} \quad = \quad -\tau$$

The fast phase ends when $\tau \nearrow \infty$ where

$$x_0 \sim -2 + 3e^{-3\tau-1}$$

At $\mu^{-4/3}$: $\qquad x_1'' + x_1'(x_0^2 - 1) + 2x_0'x_1x_0 = 0$

Integrating and matching backwards to $x_1 \sim -\tfrac{1}{3}s_0\tau$ as $\tau \searrow -\infty$

$$x_1' + x_1x_0^2 - x_1 = -s_0$$

This linear equation can be solved, but we only need the form as $\tau \nearrow \infty$. In this limit the equation becomes

$$x_1' + 3x_1 = -s_0$$

and so $x_1 \sim -\tfrac{1}{3}s_0$ as $\tau \nearrow \infty$.

At μ^{-2}: $\qquad x_2'' + x_2'(x_0^2 - 1) + 2x_0'x_2x_0 = -x_0$

Again we need only the form of the solution as $\tau \nearrow \infty$. In this limit the equation becomes

$$x_2'' + 3x_2' = 2$$

and so $x_2 \sim \tfrac{2}{3}\tau$ as $\tau \nearrow \infty$.

Matching forwards we reconstruct our solution in the fast phase in terms of the original time variable as it approaches its breakdown as $\tau \nearrow \infty$:

$$x \sim [-2 + \cdots] + \mu^{-4/3}\left[-\tfrac{1}{3}s_0 + \cdots\right]$$
$$+ \mu^{-2}\left[\tfrac{2}{3}\mu(t - \mu^{-1/3}s_0) + \cdots\right] + \cdots$$

The asymptoticness is thus broken when $t - \tfrac{3}{2}\mu^{-1/3}s_0 = \operatorname{ord}(\mu)$; broken by the particular integral in the correction term (the effect of the restoring force, neglected in the lowest approximation of the fast phase).

Second slow phase. Following the fast phase there is a repetition of the slow phase with x reversed in sign and time shifted by half the period, $\tfrac{1}{2}T$. At the lowest approximation we therefore have a solution

$$T - \mu^{-1}\tfrac{1}{2}T = \ln(-X_0) - \tfrac{1}{2}(X_0^2 - 1)$$

Matching backwards into the fast phase which ends near $x = -2$, this second slow phase solution has

$$X_0 \sim -2 + \tfrac{2}{3}\left[T - \mu^{-1}\tfrac{1}{2}T - \ln 2 + \tfrac{3}{2}\right]$$

So comparing with the end of the fast region, we see that the period of the relaxation oscillator is

$$T \sim \mu(3 - 2\ln 2) + 3\mu^{-1/3}s_0 \qquad \text{as } \mu \to \infty$$

The first term in the period comes from the slow phase, and in some sense $2\mu^{-1/3}s_0$ comes from a delay in the transition phase, while the other $\mu^{-1/3}s_0$ comes from overshooting $x = -2$ slightly.

From our solution we can see that the maximum displacement of the relaxation oscillation is at the end of the fast phase near $x = -2$:

$$\max|x| \sim 2 + \mu^{-4/3}\tfrac{1}{3}s_0 \qquad \text{as } \mu \to \infty$$

And the maximum velocity occurs in the fast phase near $x = -1$:

$$\max|\dot{x}| \sim \tfrac{4}{3}\mu + \mu^{-1/3}s_0 \qquad \text{as } \mu \to \infty$$

Higher order terms in the expansion can be obtained with some effort. The next term is just larger than the indicated μ^{-2} – it is $\mu^{-2}\ln\mu$ followed by μ^{-2}.

6

Method of strained co-ordinates

This method is due to Lighthill in 1949. It therefore predates the method of matched asymptotic expansions, and it is more limited.

The method assumes that the function to be calculated $f(x, \epsilon)$ has the same form as the unperturbed function $f(x, 0)$, but the x-scale is slightly shifted and/or slightly distorted, i.e. a strained co-ordinate. A small strain can make an asymptotic expansion for $f(x, \epsilon)$ non-uniform in x.

- An example with a small shift is

$$f(x, \epsilon) \equiv \frac{1}{x + \epsilon} \sim \frac{1}{x} - \frac{\epsilon}{x^2} + \frac{\epsilon^2}{x^3} \qquad \text{as } \epsilon \to 0$$

This is not uniformly asymptotic: there is trouble at $x = \text{ord}(\epsilon)$ as $\epsilon \to 0$.

- An example with a small distortion is

$$f(x, \epsilon) \equiv \sin(1 + \epsilon)x \sim \sin x + \epsilon x \cos x$$
$$- \tfrac{1}{2}\epsilon^2 x^2 \sin x \qquad \text{as } \epsilon \to 0$$

This is not uniformly asymptotic: there is trouble at $x = \text{ord}(\epsilon^{-1})$ as $\epsilon \to 0$.

In both these examples, the leading order term is essentially the correct form: it just needs a slight shift or distortion to make it uniformly asymptotic.

In solving differential equations we shall seek a near identity transformation $x(s, \epsilon)$ which makes $f(x, 0)$ uniformly asymptotic to $f(x, \epsilon)$. Thus we pose

$$f(x, \epsilon) \sim f_0(s) + \epsilon f_1(s)$$
$$x(s, \epsilon) \sim s + \epsilon x_1(s)$$

with the constraint that the expansion for $f(x, \epsilon)$ (but usually not for $x(s, \epsilon)$) is uniformly asymptotic. The excessive freedom in looking for the two functions in place of one is controlled by keeping everything simple in some appropriate sense (quite subjective). Here we are essentially caught by the non-uniqueness of non-Poincaré expansions.

Before applying the method of strained co-ordinates one needs some reason to believe that $f(x, \epsilon)$ is like $f(x, 0)$: the method will not work when this is not true, such as in boundary layer problems.

6.1   A model problem

In the model problem of Carrier we seek $f(x, \epsilon)$ as $\epsilon \searrow 0$ where $f(x, \epsilon)$ satisfies

$$(x + \epsilon f)f' + f = 1 \quad \text{in } 0 \leq x \leq 1$$

$$\text{subject to } f = 2 \quad \text{at } x = 1$$

We can expect trouble near $x = 0$ where x fails to be a good approximation to $(x + \epsilon f)$.

We start by treating the problem naively as a regular perturbation problem. When the method of strained co-ordinates is used, this naive expansion is always useful. A quick calculation yields

$$f(x, \epsilon) \quad \sim \quad \frac{1 + x}{x} + \epsilon \frac{3x^2 - 2x - 1}{2x^3}$$

The expansion is asymptotic at fixed x as $\epsilon \searrow 0$, but fails to be asymptotic at $x = \text{ord}(\epsilon^{1/2})$.

The exact solution can be obtained by integrating once and solving a quadratic for f:

$$f(x, \epsilon) \quad = \quad -\frac{x}{\epsilon} + \left(\frac{x^2}{\epsilon^2} + \frac{2(x + 1)}{\epsilon} + 4 \right)^{1/2}$$

Expanding for x fixed as $\epsilon \searrow 0$ gives the above naive expansion.

Comparing the graph of the exact solution with that of our naive expansion, we see in figure 6.1 that the shapes are similar, but that the exact solution has a 'singularity' at $x = -\epsilon f$ whereas all the approximations have singularities at $x = 0$. All we need to do is to 'blow over' the singularities of the approximations. This discussion suggests that the problem is one suitable for the method of strained co-ordinates.

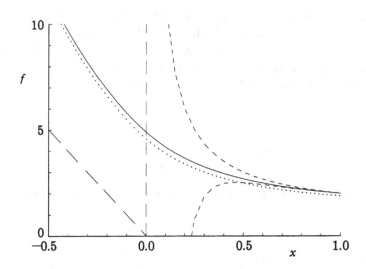

Fig. 6.1   The continuous curve gives the exact solution for $\epsilon = 0.1$. The short dashed curves on the right give the naive approximation at leading order (upper curve) and with the ϵ-correction term (lower curve). The long dashed curves are the singular lines $x = -\epsilon f$ for the exact solution and $x = 0$ for the naive approximations. The dotted curve is the uniformly asymptotic approximation corresponding to the *simplest choice* for x_1 in the method of strained co-ordinates.

6.1.1   Solution by strained co-ordinates

We pose two expansions

$$f(x,\epsilon) \quad \sim \quad f_0(s) \quad + \quad \epsilon f_1(s)$$
$$x(s,\epsilon) \quad \sim \quad s \quad + \quad \epsilon x_1(s)$$

with f_0 to be uniformly asymptotic over $0 \le x \le 1$, which is roughly $0 \le s \le 1$ but not precisely so.

In order to substitute into the governing equation, we need the change of variable in the differentiation,

$$\frac{d}{dx} \quad = \quad \frac{ds}{dx}\frac{d}{ds} \quad = \quad (1 - \epsilon x_1' + \cdots)\frac{d}{ds}$$

Now substituting into the equation and the boundary condition and comparing coefficients of ϵ^n in the usual way, we find

at ϵ^0: $sf_0' + f_0 = 1$ with $f_0 = 2$ at $s = 1$

Thus we recover the leading order result of the naive approximation,

$$f_0 = \frac{1+s}{s}$$

At ϵ^1:
$$sf_1' + f_1 = -f_0'(x_1 + f_0 - sx_1')$$
$$= \frac{x_1}{s^2} - \frac{x_1'}{s} + \frac{1}{s^3} + \frac{1}{s^2}$$

An unusual feature of this model problem is that we can write down the general solution to the above equation for f_1. In a normal problem one does not need the general solution.

$$f_1 = \frac{A}{s} - \frac{x_1}{s^2} - \frac{1}{2s^3} - \frac{1}{s^2}$$

where A is the constant of integration. Now the boundary condition is to be applied at $x = 1$, i.e. at $s = 1 - \epsilon x_1(1) + O(\epsilon^2)$. This gives the condition

$$f_0(1) + \epsilon[-x_1(1)f_0'(1) + f_1(1)] + O(\epsilon^2) = 2$$

Therefore $f_1(1) = -x_1(1)$, and so making our general solution satisfy the boundary condition we have

$$f_1 = -\frac{x_1}{s^2} + \frac{3s^2 - 2s - 1}{2s^3}$$

We now must select x_1 so that $f_0(s)$ is uniformly asymptotic to $f(x,\epsilon)$ on $0 \le x \le 1$. Clearly we should remove the s^{-3} and s^{-2} terms from f_1 which are more singular than f_0. Thus the *simplest choice* is

$$x_1 = -\frac{1+2s}{2s} \quad \text{with} \quad f_1(s) = \frac{3}{2s}$$

With this choice for x_1, one can invert $x = s + \epsilon x_1(s)$ for an approximation to $s(x,\epsilon)$ which can then be substituted into f_0. This yields the uniformly asymptotic expansion

$$f(x,\epsilon) \sim -\frac{x}{\epsilon} + \left(\frac{x^2}{\epsilon^2} + \frac{2(x+1)}{\epsilon} + 1\right)^{1/2}$$

Comparing this expression with the exact solution, we see that the above approximation has a relative error ord(ϵ) uniformly on $0 \le x \le 1$. Adding in the $\epsilon f_1(s)$ reduces the relative error to ord(ϵ^2) for x fixed as $\epsilon \searrow 0$, but leaves the relative error as ord(ϵ) for $x = $ ord($\epsilon^{1/2}$). To reduce the error when $x = $ ord($\epsilon^{1/2}$), we need to include $\epsilon^2 x_2(s)$.

An *alternative choice* for $x_1(s)$ is one that vanishes at the boundary, and consequentially leads to no forcing of $f_1(s)$ from the boundary condition.

$$x_1(s) = \frac{3s^2 - 2s - 1}{2s} \quad \text{with} \quad f_1(x) \equiv 0$$

Due to the simplicity of the model problem, this choice gives the exact solution.

Because we only have to make $f_0(s)$ uniformly asymptotic to $f(x,\epsilon)$ in $0 \le x \le 1$, which is not strictly $0 \le s \le 1$ but rather $\mathrm{ord}(\epsilon^{1/2}) \le s \le 1 + \mathrm{ord}(\epsilon)$, we can allow $f_1(s)$ to be more singular than $f_0(s)$. Thus a *third choice* for x_1 just takes out the worst singularity in f_1:

$$x_1(s) = -\frac{1}{2s} \quad \text{with} \quad f_1(2) = \frac{3s-2}{2s^2}$$

This choice gives a fractional error $\mathrm{ord}(\epsilon^{1/2})$ when $x = \mathrm{ord}(\epsilon^{1/2})$, worse because of the dangerous choice of f_1.

Note there is no need to calculate $f_1(s)$ if only the leading order uniformly asymptotic expansion, $f_0(s)$ with $x = s + \epsilon x_1(s)$, is required. One can choose an x_1 which removes the forcing of that part of f_1 which would be more singular than f_0 (the first choice above), or one can choose the x_1 which makes f_1 vanish identically (the second choice above), although the calculation of this x_1 involves more work.

6.1.2 Solution by method of matched asymptotic expansions

The naive expansion serves as an outer approximation. It breaks down at $x = \mathrm{ord}(\epsilon^{1/2})$ where $f = \mathrm{ord}(\epsilon^{-1/2})$. This suggests a rescaling

$$x = \epsilon^{1/2}\xi$$

for the inner approximation and an expansion in powers of $\epsilon^{1/2}$ starting at $\epsilon^{-1/2}$, i.e.

$$f(x,\epsilon) \sim \epsilon^{-1/2}F_0(\xi) + F_1(\xi) + \epsilon^{1/2}F_2(\xi)$$

Substituting into the governing equation and comparing coefficients of $\epsilon^{n/2}$ in the usual way we find the following.

At $\epsilon^{-1/2}$: $(\xi + F_0)F_0' + F_0 = 0$

The general positive solution is

$$F_0 = (\xi^2 + A_0)^{1/2} - \xi$$

with A_0 a constant of integration.

At ϵ^0: $\quad\quad \xi F_1' + F_1 F_0' + F_0 F_1' + F_1 = 1$

This linear equation has a general solution

$$F_1 = \frac{\xi + A_1}{(\xi^2 + A_0)^{1/2}}$$

with A_1 a constant of integration.

At $\epsilon^{1/2}$: $\quad\quad \xi F_2' + F_2 F_0' + F_1 F_1' + F_0 F_2' + F_2 = 0$

This linear equation has a general solution

$$F_2 = \frac{A_2 - \frac{1}{2} F_1^2}{(\xi^2 + A_0)^{1/2}}$$

with A_2 a constant of integration.

Matching with an intermediate variable $\eta = x/\epsilon^\alpha = \epsilon^{1/2-\alpha}\xi$ fixed as $\epsilon \searrow 0$ with $0 < \alpha < \frac{1}{2}$, we have

$$\text{outer} \;=\; \frac{\epsilon^{-\alpha}}{\eta} + 1 + \epsilon\left(-\frac{\epsilon^{-3\alpha}}{2\eta^3} - \frac{\epsilon^{-2\alpha}}{\eta^2} + \frac{3\epsilon^{-\alpha}}{2\eta} \right) + \cdots$$

$$\text{inner} \;=\; \epsilon^{-1/2}\left(\frac{A_0 \epsilon^{1/2-\alpha}}{2\eta} - \frac{A_0^2 \epsilon^{3/2-3\alpha}}{8\eta^3} + \cdots \right)$$

$$+ \left(1 + \frac{A_1 \epsilon^{1/2-\alpha}}{\eta} - \frac{A_0 \epsilon^{1-2\alpha}}{2\eta^2} + \cdots \right)$$

$$+ \epsilon^{1/2}\left(\frac{(A_2 - \frac{1}{2})\epsilon^{1/2-\alpha}}{\eta} + \cdots \right) + \cdots$$

Thus

at $\epsilon^{-\alpha}$: $\quad\quad \dfrac{1}{\eta} = \dfrac{A_0}{2\eta},$ $\quad\quad$ i.e $A_0 = 2$

at ϵ^0: $\quad\quad 1 = 1$

at $\epsilon^{1/2-\alpha}$: $\quad\quad 0 = \dfrac{A_1}{\eta},$ $\quad\quad$ i.e. $A_1 = 0$

at $\epsilon^{1-3\alpha}$: $\quad\quad -\dfrac{1}{2\eta^3} = -\dfrac{A_0^2}{8\eta^3},$ $\quad\quad$ i.e. $A_0 = 2$ again

at $\epsilon^{1-2\alpha}$: $\quad\quad -\dfrac{1}{\eta^2} = -\dfrac{A_0}{2\eta^2},$ $\quad\quad$ i.e. $A_0 = 2$ again

at $\epsilon^{1-\alpha}$: $\quad\quad \dfrac{3}{2\eta} = (A_2 - \frac{1}{2})\dfrac{1}{\eta},$ $\quad\quad$ i.e. $A_2 = 2$

This completes the solution by matched asymptotic expansions.

Comparing the two methods of solution, we see that the method of strained co-ordinates produces a uniformly asymptotic approximation faster – we still have a little work to form the uniformly asymptotic composite approximation in the above matched asymptotic solution. On the other hand one does have a clearer idea about the form of the errors with the matched asymptotic expansions – it was not apparent that there were $\epsilon^{1/2}$ errors from the strained co-ordinate solution. Also at higher orders the matched asymptotic expansion method is more routine. Finally one must not forget that the strained co-ordinate method is limited to problems which have $f(x,0)$ of the same shape as $f(x,\epsilon)$.

Exercise 6.1. (Carrier)   The function $f(x,\epsilon)$ satisfies the equation

$$(x^2 + \epsilon f)f' + f = 0 \quad \text{in } 0 \le x \le 1$$

and is subject to the boundary condition $f = e$ at $x = 1$. Find f at $x = 0$.

Exercise 6.2.   The function $f(x,\epsilon)$ satisfies the equation

$$(x + \epsilon f)f' - \tfrac{1}{2}f = 1 + \tfrac{1}{2}\epsilon$$

and is subject to the boundary condition $f = -2 + \sqrt{1 - 2\epsilon}$ at $x = 1$. Find $f(x,\epsilon)$. Note that the exact solution is $f = -2 + \sqrt{x - 2\epsilon}$ which is real for $x > 2\epsilon$.

Exercise 6.3. (Carrier)   The function $v(r,z;\epsilon)$ satisfies

$$\frac{\partial v}{\partial r} + \frac{v}{r} + v\frac{\partial v}{\partial z} = 0 \quad \text{in } r \ge 1$$

and is subject to the boundary condition

$$v = \epsilon f(z) \quad \text{on } r = 1$$

where $f(z)$ is a given function. Find the non-uniformity in the naive approximation, and then solve by straining the r-coordinate seeking a solution in the form $v(s,z;\epsilon)$ and $r(s,z;\epsilon)$.

6.2   Duffing's oscillator

This is an oscillator with a slightly nonlinear restoring force. We shall solve for $x(t,\epsilon)$ satisfying

$$\ddot{x} + x - \epsilon x^3 = 0 \quad \text{in } t \ge 0$$

and subject to initial conditions

$$x = 1 \quad \text{and} \quad \dot{x} = 0 \quad \text{at } t = 0$$

Treating the problem naively as a regular one we obtain the approximation

$$x(t, \epsilon) \quad \sim \quad \cos t + \epsilon \left(\tfrac{3}{8} t \sin t + \tfrac{1}{32} (\cos t - \cos 3t) \right)$$

This is asymptotic for fixed t as $\epsilon \to 0$, but loses its asymptoticness when $t \geq \mathrm{ord}(\epsilon)$.

Now the equation describes a conservative system. There will therefore be an oscillation with a constant amplitude. The trouble is that the period of the oscillation is slightly different from 2π. The problem is thus one suitable for analysis by the method of strained co-ordinates.

We pose the two expansions

$$\begin{aligned}
x(t, \epsilon) &\sim & x_0(s) &+& \epsilon x_1(s) &+& \epsilon^2 x_2(s) \\
t(s, \epsilon) &\sim & s &+& \epsilon t_1(s) &+& \epsilon^2 t_2(s)
\end{aligned}$$

Substituting into the governing equation and boundary conditions, and comparing coefficients of ϵ^n, we find a sequence of problems.

At ϵ^0: $\qquad x_0'' + x_0 = 0 \quad$ with $x_0 = 1$ and $x_0' = 0$ at $s = 0$

Thus $x_0(s) = \cos s$ as in the naive approximation.

At ϵ^1:
$$\begin{aligned}
x_1'' + x_1 &= x_0^3 + 2t_1' x_0'' + t_1'' x_0' \\
&= \left(\tfrac{3}{4} - 2t_1' \right) \cos s - t_1'' \sin s + \tfrac{1}{4} \cos 3s
\end{aligned}$$

Now in order to keep $x_0(s)$ uniformly asymptotic, we make x_1 bounded and hence require the resonant forcing of the simple harmonic oscillator to vanish, i.e.

$$\tfrac{3}{4} - 2t_1' = 0 \qquad \text{and} \qquad t_1'' = 0$$

For the initial conditions at this order we choose to make the straining vanish at the initial instant, which leads to the simplified conditions

$$x_1 = x_1' = t_1 = 0 \qquad \text{at } s = 0$$

Hence

$$t_1(s) = \tfrac{3}{8} s \qquad \text{and} \qquad x_1(s) = \tfrac{1}{32} (\cos s - \cos 3s)$$

At ϵ^2:
$$\begin{aligned}
x_2'' + x_2 &= 3x_0^2 x_1 + 2t_1' x_1'' + t_1'' x_1' \\
&\quad + (2t_2' - 3t_1'^2) x_0'' + (t_2'' - 3t_1' t_1'') x_0' \\
&= \left(\tfrac{57}{128} - 2t_2' \right) \cos s - t_2'' \sin s + \tfrac{3}{16} \cos 3s - \tfrac{3}{128} \cos 5s
\end{aligned}$$

Again removing the resonant terms, and satisfying initial conditions with no straining at the initial instant, i.e. $x_2 = x_2' = t_2 = 0$ at $s = 0$, we find

$$t_2 = \tfrac{57}{256}s \qquad \text{and} \qquad x_2 = \tfrac{1}{1024}(23\cos s - 24\cos 3s + \cos 5s)$$

We now have calculated the change in the period of the nonlinear oscillator to be

$$2\pi \left(1 + \tfrac{3}{8}\epsilon + \tfrac{57}{256}\epsilon^2 + \cdots\right)$$

6.3   Shallow water waves

For waves with a wavelength much greater than the depth of water, it can be shown that the velocity is approximately horizontal and nearly independent of depth and that with negligible vertical accelerations the pressure is approximately hydrostatic. The depth integrated equations of motion are then

$$\text{Mass Conservation:} \quad \frac{\partial h}{\partial t} + u\frac{\partial h}{\partial x} + h\frac{\partial u}{\partial x} = 0$$

$$\text{Momentum Conservation:} \quad \frac{\partial u}{\partial t} + u\frac{\partial u}{\partial x} + \frac{\partial h}{\partial x} = 0$$

These nonlinear equations propagate waves in both directions. We will consider nearly linear waves with an initial disturbance

$$u = \epsilon\sin x \qquad \text{and} \qquad h = 1 + \epsilon\sin x + \tfrac{1}{4}\epsilon^2\sin^2 x \qquad \text{at } t = 0$$

which have been chosen so that waves are created moving in only one direction.

If we naively treat the problem as a regular one, we find

$$u(x,t;\epsilon) \sim 0 + \epsilon\sin(x - t) - \epsilon^2\tfrac{3}{4}t\sin 2(x - t)$$
$$h(x,t;\epsilon) \sim 1 + \epsilon\sin(x - t) + \epsilon^2\left(-\tfrac{3}{4}t\sin 2(x - t) + \tfrac{1}{4}\sin^2(x - t)\right)$$

This naive approximation is asymptotic for fixed t as $\epsilon \to 0$, but loses its asymptoticness at $t \geq \text{ord}(\epsilon^{-1})$. The trouble is that each approximation has characteristics $x \pm t = constant$, whereas the nonlinear equation has slightly different characteristics. The travelling wave is basically correct: the naive approximation breaks down because it positions the wave wrongly after a long time. This problem is therefore suited to co-ordinate straining, and the strained co-ordinates are called the characteristics!

We start by converting the governing equations into their characteristic form $Q_t + cQ_x = 0$ for some suitable quantity Q being propagated

at speed c. Adding the mass equation to the momentum equation multiplying by λ, we have

$$[u_t + \lambda h_t] + [(u + \lambda h)u_x + (1 + \lambda u)h_x] = 0$$

Requiring that the ratio of the time derivatives of u and h be equal to
the ratio of the space derivatives gives

$$\frac{1}{\lambda} = \frac{u + \lambda h}{1 + \lambda u}$$

Hence $\lambda = \pm h^{-1/2}$, so we produce a pair of characteristic equations

$$\left(\frac{\partial}{\partial t} + (u \pm h^{1/2})\frac{\partial}{\partial x}\right)\left(u \pm 2h^{1/2}\right) = 0$$

We now transform to new independent variables α and β, the characteristics. Thus we have $x(\alpha, \beta)$ and $t(\alpha, \beta)$, so that

$$\left(\frac{\partial}{\partial \alpha}\right)_\beta = \left(\frac{\partial t}{\partial \alpha}\right)_\beta \left(\frac{\partial}{\partial t}\right)_x + \left(\frac{\partial x}{\partial \alpha}\right)_\beta \left(\frac{\partial}{\partial x}\right)_t$$

Comparing this with the new pair of governing equations, we see that if
we take

$$\left(\frac{\partial x}{\partial \alpha}\right)_\beta = (u + h^{1/2})\left(\frac{\partial t}{\partial \alpha}\right)_\beta \quad \text{then} \quad \left(\frac{\partial}{\partial \alpha}\right)_\beta \left(u + 2h^{1/2}\right) = 0$$

and

$$\left(\frac{\partial x}{\partial \beta}\right)_\alpha = (u - h^{1/2})\left(\frac{\partial t}{\partial \beta}\right)_\alpha \quad \text{then} \quad \left(\frac{\partial}{\partial \beta}\right)_\alpha \left(u - 2h^{1/2}\right) = 0$$

The two equations on the right can be integrated to yield the Riemann
invariants

$$u + 2h^{1/2} = \text{function only of } \beta$$

and

$$u - 2h^{1/2} = \text{function only of } \alpha$$

Before applying the initial conditions, we need to select a normalisation of the labelling of the characteristics. The differential equations
just describe how the values change away from the initial condition. The
most convenient labelling is

$$\alpha = \beta = x \quad \text{on } t = 0$$

Now we can evaluate the two conserved Riemann invariants by moving back along their characteristics to the initial data.

$$
\begin{aligned}
u + 2h^{1/2} &= \quad \text{function of } \beta \\
&= \quad \epsilon \sin \beta + 2(1 + \tfrac{1}{2}\epsilon \sin \beta) \qquad \text{from } t = 0 \text{ where } x = \beta \\
&= \quad 2 + 2\epsilon \sin \beta \\
u - 2h^{1/2} &= \quad \text{function of } \alpha \\
&= \quad \epsilon \sin \alpha - 2(1 + \tfrac{1}{2}\epsilon \sin \alpha) \qquad \text{from } t = 0 \text{ where } x = \alpha \\
&= \quad -2
\end{aligned}
$$

In this model problem it is not necessary to expand the initial data in powers of ϵ. We now have the exact solution for u and h in terms of the strained co-ordinates.

$$
\begin{aligned}
u(\alpha, \beta) &= \quad \epsilon \sin \beta \\
h(\alpha, \beta) &= \quad 1 + \epsilon \sin \beta + \tfrac{1}{4}\epsilon^2 \sin^2 \beta
\end{aligned}
$$

The solution is independent of α because the initial data was chosen to have waves propagating only in one direction.

There now remains the problem of relating the values of the characteristics to the original variables x and t. From our exact solution we

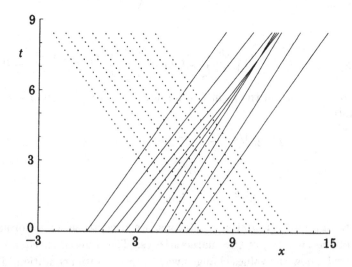

Fig. 6.2　The characteristics on the xt-plane for $\epsilon = 0.1$. The dotted curves are the characteristics with $\alpha = \text{const}$ while the continuous curves are the characteristics with $\beta = \text{const}$.

can evaluate the slopes of the characteristics.

$$\left(\frac{\partial x}{\partial t}\right)_\beta = \left(\frac{\partial x}{\partial \alpha}\right)_\beta \bigg/ \left(\frac{\partial t}{\partial \alpha}\right)_\beta = u + h^{1/2} = 1 + \tfrac{3}{2}\epsilon \sin \beta$$

$$\left(\frac{\partial x}{\partial t}\right)_\alpha = \left(\frac{\partial x}{\partial \beta}\right)_\alpha \bigg/ \left(\frac{\partial t}{\partial \beta}\right)_\alpha = u - h^{1/2} = -1 + \tfrac{1}{2}\epsilon \sin \beta$$

Unfortunately the equations for the characteristics cannot be solved exactly, and so we must seek an expansion of the co-ordinate straining in powers of ϵ.

$$x(\alpha, \beta; \epsilon) \sim x_0(\alpha, \beta) + \epsilon x_1(\alpha, \beta) + \epsilon^2 x_2(\alpha, \beta)$$
$$t(\alpha, \beta; \epsilon) \sim t_0(\alpha, \beta) + \epsilon t_1(\alpha, \beta) + \epsilon^2 t_2(\alpha, \beta)$$

Substituting into the above equations for the slopes of the characteristics and comparing coefficients of ϵ^n, we find a sequence of problems.

At ϵ^0: for the characteristic with $\beta = $ constant

$$\left(\frac{\partial x_0}{\partial \alpha}\right)_\beta = \left(\frac{\partial t_0}{\partial \alpha}\right)_\beta$$

This has a solution $x_0 = t_0 + $ function of β. The unknown function of β is determined by referring to our choice of the labelling of the characteristics, i.e. $\alpha = \beta = x$ at $t = 0$. Thus

$$x_0 = t_0 + \beta$$

Similarly for the α-characteristic. Then solving for x_0 and t_0 we obtain

$$x_0 = \tfrac{1}{2}(\alpha + \beta) \qquad \text{and} \qquad t_0 = \tfrac{1}{2}(\alpha - \beta)$$

At ϵ^1: for the characteristic with $\beta = $ constant

$$\left(\frac{\partial x_1}{\partial \alpha}\right)_\beta = \left(\frac{\partial t_1}{\partial \alpha}\right)_\beta + \tfrac{3}{2}\sin \beta \left(\frac{\partial t_0}{\partial \alpha}\right)_\beta = \left(\frac{\partial t_1}{\partial \alpha}\right)_\beta + \tfrac{3}{4}\sin \beta$$

This has a solution satisfying the labelling $x_1 = t_1 = 0$ on $\alpha = \beta = x$

$$x_1 = t_1 + \tfrac{3}{4}(\alpha - \beta)\sin \beta$$

Similarly for the α-characteristic. Then solving for x_1 and t_1 we obtain

$$x_1 = \tfrac{3}{8}(\alpha - \beta)\sin \beta + \tfrac{1}{8}(\cos \beta - \cos \alpha)$$
$$t_1 = -\tfrac{3}{8}(\alpha - \beta)\sin \beta + \tfrac{1}{8}(\cos \beta - \cos \alpha)$$

At ϵ^2: repeating the above process produces

$$x_2 = \tfrac{1}{128}\big[(\alpha - \beta)(-22 + 21\cos 2\beta)$$
$$+ (11\sin \beta \cos \beta + \sin \alpha \cos \alpha - 12\sin \beta \cos \alpha)\big]$$

$$t_2 \quad = \quad \tfrac{1}{128}\big[(\alpha - \beta)(14 - 15\cos 2\beta)$$
$$+ (-13\sin\beta\cos\beta + \sin\alpha\cos\alpha + 12\sin\beta\cos\alpha)\big]$$

Note that the expansion is uniformly asymptotic, i.e. each t_n has only the factor $(\alpha - \beta)^1$ and no higher power. The solution is therefore valid into the region $t \geq \mathrm{ord}(\epsilon^{-1})$. We can therefore study the phenomenon of wave breaking.

A wave breaks when there is more than one value for h for one value of x and t – see figure 6.3. Now h is a unique function of α and β. The non-uniqueness must therefore come from the transformation from x and t to α and β. When the non-uniqueness first starts the Jacobian of the transformation will vanish, i.e.

$$0 \quad = \quad \left(\frac{\partial x}{\partial \alpha}\right)_\beta \left(\frac{\partial t}{\partial \beta}\right)_\alpha - \left(\frac{\partial x}{\partial \beta}\right)_\alpha \left(\frac{\partial t}{\partial \alpha}\right)_\beta$$
$$= \quad 2h^{1/2}\left(\frac{\partial t}{\partial \alpha}\right)_\beta \left(\frac{\partial t}{\partial \beta}\right)_\alpha$$

Now $h^{1/2}$ and $\partial t/\partial \alpha$ are both positive. Hence wave breaking occurs where

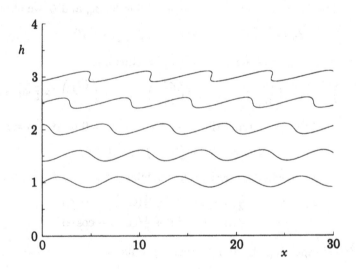

Fig. 6.3   The breaking wave for $\epsilon = 0.1$. The curves give h as a function of x at $t = 0, 2, 4, 6$ and 8 with vertical offsets of $\tfrac{1}{4}t$.

$$0 = \left(\frac{\partial t}{\partial \beta}\right)_\alpha = -\tfrac{1}{2} + \epsilon\left(-\tfrac{3}{8}(\alpha-\beta)\cos\beta + \tfrac{1}{4}\sin\beta\right)$$
$$+ \epsilon^2\left(\tfrac{15}{64}(\alpha-\beta)\sin 2\beta + O(1)\right) + \cdots$$

i.e. where

$$(\alpha - \beta) = -\frac{4}{3\epsilon\cos\beta}\left(1 + \tfrac{3}{4}\epsilon\sin\beta + O(\epsilon^2)\right)$$

i.e. when

$$t = -\frac{2}{3\epsilon\cos\beta}\left(1 + O(\epsilon^2)\right)$$

Hence the earliest time that the wave breaks is at $t = \tfrac{2}{3}\epsilon^{-1} + O(\epsilon)$ at locations $x = \tfrac{2}{3}\epsilon^{-1} + (2n+1)\pi + O(\epsilon)$. Because the initial conditions have been chosen so that waves are propagating in only one direction, it is possible to find exactly when and where the waves first break, and thus one finds that the $O(\epsilon)$ possible errors in the asymptotic analysis are actually zero.

7

Method of multiple scales

This is a general method applicable to a wide range of problems. The problems are characterised by having two physical processes, each with their own scales, and with the two processes acting simultaneously. This should be contrasted with the method of matched asymptotic expansions which also has two processes with different scales, but with the processes acting separately in different regions.

7.1   van der Pol oscillator

We return to the oscillator which was introduced in §5.6 now to consider the case where the nonlinear friction is a small perturbation to the linear simple harmonic oscillator. It is convenient to study the initial value problem

$$\ddot{x} + \epsilon \dot{x}(x^2 - 1) + x = 0 \qquad \text{in } t \geq 0 \text{ with } \epsilon \searrow 0$$
$$\text{subject to} \quad x = 1 \text{ and } \dot{x} = 0 \quad \text{at } t = 0$$

Treating the problem as a regular one yields the approximation

$$x(t, \epsilon) \quad \sim \quad \cos t + \epsilon \left[\tfrac{3}{8}(t \cos t - \sin t) - \tfrac{1}{32}(\sin 3t - 3 \sin t) \right]$$

This expansion is asymptotic for fixed t as $\epsilon \searrow 0$, but breaks down when $t \geq \text{ord}(\epsilon^{-1})$. (It is however possible to prove that the expansion converges!)

The trouble with the naive approximation is that the ϵ-damping changes the amplitude of the oscillation on a time scale ϵ^{-1} by the slow accumulation of small effects. Thus the oscillator has two processes acting on their own time scales. There is the basic oscillation on the time scale of 1 from the inertia causing the restoring force to overshoot the equilibrium position. There is also the slow drift in the amplitude (and possibly the phase) on the time scale ϵ^{-1} due to the small friction. We recognise these two time scales by introducing two time variables.

$$\tau = t \quad \text{– the } \textit{fast time} \text{ of the oscillation}$$
$$T = \epsilon t \quad \text{– the } \textit{slow time} \text{ of the amplitude drift}$$

The slowly changing features will then be combined into factors which are functions of T, while the rapidly changing features will be combined into factors which are functions of τ. Thus we look for a solution of the form

$$x(t; \epsilon) \; = \; x(\tau, T; \epsilon)$$

As real time t increases the fast time τ increases at the same rate, while the slow time T increases slowly. Thus

$$\frac{d}{dt} \; = \; \left(\frac{\partial}{\partial \tau}\right)_T + \epsilon \left(\frac{\partial}{\partial T}\right)_\tau$$

and so

$$\ddot{x} \; = \; x_{\tau\tau} + 2\epsilon x_{\tau T} + \epsilon^2 x_{TT}$$

We now seek an asymptotic approximation for x allowing in the leading order for the possibility of changes over the long time scale. Thus we pose

$$x(t; \epsilon) \quad \sim \quad x_0(\tau, T) + \epsilon x_1(\tau, T)$$

with the requirement that the expansion be asymptotic for $T = \text{ord}(1)$. Substituting into the governing equation and comparing coefficients of ϵ^n, we find a sequence of problems.

At ϵ^0:

$$x_{0\tau\tau} + x_0 = 0 \qquad \text{in } t \geq 0$$
$$\text{with} \quad x_0 = 1 \quad \text{and} \quad x_{0\tau} = 0 \quad \text{at } t = 0$$

Integrating with respect to τ, treating T as an independent variable held constant, we obtain a general solution

$$x_0 \; = \; R(T) \cos\left(\tau + \theta(T)\right)$$

in which the amplitude R and the phase θ are constant as far as the rapid τ variations are concerned, but are allowed to vary over the long T time. The initial conditions give

$$R(0) \; = \; 1 \qquad \text{and} \qquad \theta(0) \; = \; 0$$

Except for this information, R and θ are unknown in the leading order analysis. Knowing that the amplitude is controlled by the action of the small friction over a long time, it is quite clear that we must proceed to the next order.

At ϵ^1:

$$
\begin{aligned}
x_{1\tau\tau} + x_1 &= -x_{0\tau}(x_0^2 - 1) - 2x_{0\tau T} \quad \text{in } t \geq 0 \\
&= 2R\theta_T \cos(\tau + \theta) + \left(2R_T + \tfrac{1}{4}R^3 - R\right)\sin(\tau + \theta) \\
&\quad + \tfrac{1}{4}R^3 \sin 3(\tau + \theta)
\end{aligned}
$$

from the (partially) known x_0. The initial conditions are

$$
x_1 = 0 \quad \text{and} \quad x_{1\tau} = -x_{0T} = -R_T \quad \text{at } t = 0
$$

Again we integrate with respect to τ treating T as a constant. The $\sin 3(\tau + \theta)$ forcing term will induce a $\sin 3(\tau + \theta)$ bounded response in x_1, but the resonating forcing terms $\sin(\tau + \theta)$ and $\cos(\tau + \theta)$ would induce a response in x_1 growing like τ which would break the asymptoticness when $t \geq \text{ord}(\epsilon^{-1})$. Thus to maintain the asymptoticness of the expansion we must exploit the freedom in the undetermined $R(T)$ and $\theta(T)$ in order to insist that the potentially resonating terms vanish identically.   This leads to the so-called secularity or integrability or solubility condition of Poincaré,

$$
\theta_T = 0 \quad \text{and} \quad R_T = \tfrac{1}{8}R\left(4 - R^2\right)
$$

Using the initial conditions on R and θ, we therefore have

$$
\theta \equiv 0 \quad \text{and} \quad R = 2\left(1 + 3e^{-T}\right)^{-1/2}
$$

Thus eventually the amplitude of the oscillator drifts to 2. Note that the amplitude and phase of the leading order term are fully determined in the next order problem by asking that the correction term does not break the asymptoticness; it is not necessary to find the correction term.

The correction term can however be found, and is

$$
x_1 = -\tfrac{1}{32}R^3 \sin 3\tau + S(T)\sin\left(\tau + \varphi(T)\right)
$$

with new unknown amplitude and phase functions, $S(T)$ and $\varphi(T)$, which satisfy the initial conditions

$$
\varphi(0) = 0 \quad \text{and} \quad S(0) = -\tfrac{9}{32}
$$

These new amplitude and phase functions will become determined by the secularity condition in the x_2 problem.

At higher orders one can find that a resonant forcing is unavoidable: there can be insufficient freedom in the undetermined functions. The asymptoticness is then lost. This is in fact the situation with our van der Pol oscillator when proceeding to find x to ord(ϵ) for t of ord(ϵ^{-1}). This difficulty can be overcome by introducing an additional slow time scale $T_2 = \epsilon^2 t$.

A simple example which illustrates the need for such a super slow time scale is a linearly damped oscillator

$$\ddot{x} + 2\epsilon\dot{x} + x = 0$$

with solution

$$x = e^{-\epsilon t}\cos\left(\sqrt{1 - \epsilon^2}\,t\right)$$

The amplitude drifts on the time scale ϵ^{-1}, while the phase drifts on the longer time scale ϵ^{-2}. Of course in this example there is not much amplitude left by the time that the phase has slipped significantly.

In general when working to ord(ϵ^k) on a time scale ord(ϵ^{k-n}) one must expect to have a hierarchy of n slow time scales. Some may be essential, representing genuinely different processes. Some however may simply be adjustments to a previous process, e.g. adjustments in the frequency, which are better tackled with something like a co-ordinate straining.

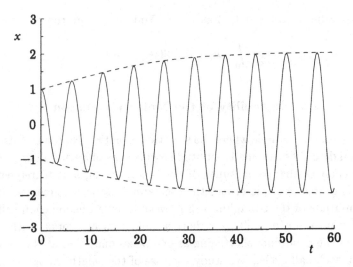

Fig. 7.1   The continuous curve is a numerical solution $x(t; \epsilon)$ of the initial value problem for a van der Pol oscillator with $\epsilon = 0.1$. The dashed curves give the asymptotic predictions for the amplitude, i.e. $\pm R(T)$. The agreement is good at even larger values of the small parameter, as is typical of this class of asymptotic analysis.

Exercise 7.1.   Obtain an asymptotic approximation for x to ord(1) which is valid for $t = \text{ord}(\epsilon^{-1})$ to the solution of

$$\ddot{x} + \epsilon \dot{x}^3 + x = 0 \qquad \text{for } t \geq 0$$
$$\text{with} \quad x = 1 \quad \text{and} \quad \dot{x} = 0 \quad \text{at } t = 0$$

Exercise 7.2.   Obtain equations for the drift in the amplitude and phase in the solution to

$$\ddot{x} + \epsilon \dot{x}(x^2 - 1) + (1 + \epsilon k)x = \epsilon \cos t$$

with $k = \text{ord}(1)$ as $\epsilon \searrow 0$. [The tough part is then to show that a slave oscillator will lock onto the forcing from a master if the slave is not detuned too much, i.e. if $|k| < k_c$ then R tends to an equilibrium, while if $|k| > k_c$ then R oscillates (the free oscillations beating with the forced response).]

Exercise 7.3.   Find the leading order approximation to the general solution for $x(t; \epsilon)$ and $y(t; \epsilon)$ satisfying

$$\frac{d^2 x}{dt^2} + 2\epsilon y \frac{dx}{dt} + x = 0$$
$$\frac{dy}{dt} = \tfrac{1}{2}\epsilon \ln x^2$$

which is valid for $t = \text{ord}(1/\epsilon)$ as $\epsilon \to 0$. You may quote the result

$$\frac{1}{2\pi} \int_0^{2\pi} \ln \cos^2 \theta \, d\theta = -\ln 4$$

7.2   Instability of the Mathieu equation

This is a simple example where the slow time is not ϵt. Now the Mathieu equation describes the small amplitude oscillations of a pendulum whose length changes slightly in time. If the length changes at a frequency which is near a multiple of half the oscillator's natural frequency, then the amplitude of the pendulum will grow in time, a phenomenon called parametric excitation. [This is how a child's swing works, with the length of the pendulum shortening as one raises one's feet at the lowest point of each half cycle.] We study the case of the length changing with roughly the same frequency of the natural oscillations:

$$\ddot{x} + (1 + k\epsilon^2 + \epsilon \cos t)x = 0$$

with $k = \text{ord}(1)$ as $\epsilon \to 0$.

Now in the leading order approximation, x will just contain the first harmonic. Iterating, this will force a correction with zero and second harmonics. Iterating again, we would find a second correction forced by first and third harmonics. The resonant forcing has to be removed by a slow drift in the amplitude and phase, which therefore occurs on a time scale $\mathrm{ord}(\epsilon^{-2})$. Hence we introduce a slow time scale $T = \epsilon^2 t$ (with the fast time $\tau = t$ unchanged) and pose an expansion

$$x(t;\epsilon) \quad \sim \quad x_0(\tau,T) + \epsilon x_1(\tau,T) + \epsilon^2 x_2(\tau,T)$$

to be asymptotic when $T = \mathrm{ord}(1)$. Substituting into the governing equation and comparing coefficients of ϵ^n, we find in the usual way a sequence of problems.

At ϵ^0:

$$x_{0\tau\tau} + x_0 = 0$$

with a general solution

$$x_0 = A(T)\cos\tau + B(T)\sin\tau$$

At ϵ^1:

$$\begin{aligned} x_{1\tau\tau} + x_1 &= -x_0 \cos\tau \\ &= -\tfrac{1}{2}A - \tfrac{1}{2}A\cos 2\tau - \tfrac{1}{2}B\sin 2\tau \end{aligned}$$

This forces a response in x_1,

$$x_1 = -\tfrac{1}{2}A + \tfrac{1}{6}A\cos 2\tau + \tfrac{1}{6}B\sin 2\tau$$

The homogeneous solution can be omitted as we are not tackling a particular initial value problem.

At ϵ^2:

$$\begin{aligned} x_{2\tau\tau} + x_2 &= -2x_{0\tau T} - kx_0 - x_1 \cos\tau \\ &= -\left(2B_T + (k - \tfrac{5}{12})A\right)\cos\tau + \left(2A_T - (k + \tfrac{1}{12})B\right)\sin\tau \\ &\quad - \tfrac{1}{12}A\cos 3\tau - \tfrac{1}{12}B\sin 3\tau \end{aligned}$$

As anticipated, the drift in amplitude and phase is determined at $\mathrm{ord}(\epsilon^2)$ by the secularity condition that the asymptoticness is not lost when $T = \mathrm{ord}(1)$,

$$B_T = \tfrac{1}{2}\left(\tfrac{5}{12} - k\right)A \qquad \text{and} \qquad A_T = \tfrac{1}{2}\left(\tfrac{1}{12} + k\right)B$$

The solution of this pair of linear equations has either exponentially growing (and decaying) or stable oscillatory solutions according to whether

$$\left(\tfrac{5}{12} - k\right)\left(\tfrac{1}{12} + k\right) \quad > \text{ or } < \quad 0$$

i.e. the oscillator is unstable if $-\tfrac{1}{12} < k < \tfrac{5}{12}$.

Exercise 7.4. The equations governing a satellite orbiting the Earth and experiencing a small frictional force proportional to the square of its velocity may be written in the form

$$u_{\theta\theta} + u = h^2$$
$$h_\theta = \epsilon h \frac{\sqrt{u_\theta^2 + u^2}}{u^2}$$

Writing the leading order solution

$$u(\theta, \epsilon) \sim h^2(\epsilon\theta)\left[1 + e(\epsilon\theta)\cos(\theta - \alpha(\epsilon\theta))\right]$$

with $e < 1$, obtain the drift equation

$$h' = \langle f \rangle / h$$
$$e' = -2\langle (e + \cos\varphi)f \rangle / h^2$$
$$\alpha' = -2\langle \sin\varphi \, f \rangle / h^2 e$$

where $f = \sqrt{1 + 2e\cos\varphi + e^2}/(1 + e\cos\varphi)^2$ and the angle brackets denote an average over $0 \le \varphi \le 2\pi$.

Deduce that as the satellite falls to the Earth ($r = 1/u$), its angular momentum h increases, its eccentricity e decreases, and the direction of the perihelion α does not drift at this order.

If the eccentricity e is small initially, find an approximate solution for the drift in h and e.

[You may assume that $\langle (e + \cos\varphi)f \rangle > 0$ for $0 < e < 1$.]

Exercise 7.5. Find the leading order approximation valid for times t of order ϵ^{-1} as $\epsilon \to 0$, to the solution $x(t; \epsilon)$ and $y(t; \epsilon)$ satisfying

$$\ddot{x} + \epsilon y \dot{x} + x = y^2$$
$$\dot{y} = \epsilon(1 + x - y - y^2)$$

subject to $x = 1$, $\dot{x} = 0$ and $y = 0$ at $t = 0$.

Exercise 7.6. Find the leading order approximation which is valid for times $t = \mathrm{ord}(\epsilon^{-1})$ as $\epsilon \to 0$, to the solution $x(t;\epsilon)$ and $y(t;\epsilon)$ satisfying

$$\frac{dx}{dt} + x^2 y \cos t = \epsilon(x - 2x^2)$$

$$\frac{dy}{dt} = \epsilon\left(1 - \frac{\sin t}{x}\right)$$

with $x = 1$ and $y = 0$ at $t = 0$.

7.3 A diffusion–advection equation

An essential feature of problems requiring the method of multiple scales is that there is some quantity which is preserved at leading order, but which can drift through the accumulation of small disturbances. In the first two sections, we had a conservative oscillator which preserved the amplitude at leading order. Over a long time, however, the amplitude drifted through respectively the accumulated effect of the ϵ damping term and the ϵ work done on the oscillator by changing the length of the pendulum. Our next application of the method of multiple scales is to a partial differential equation describing advection around a periodic domain $0 < \theta < 2\pi$ with some small diffusion. Thus at leading order information is preserved as it is advected around. While over a long time this information changes from the initial data through the accumulated effect of weak diffusion.

We consider the initial value problem for $f(\theta, t; \epsilon)$ with $\epsilon \searrow 0$

$$\frac{\partial f}{\partial t} + \frac{\partial}{\partial \theta}(\omega(\theta)f) = \epsilon\frac{\partial^2 f}{\partial \theta^2} \qquad \text{in } t \geq 0 \text{ and } 0 \leq \theta \leq 2\pi$$

$$\text{with} \quad f = F(\theta) \qquad \text{at } t = 0$$

where $\omega(\theta)$ is given over $0 \leq \theta \leq 2\pi$, periodic and positive.

The two processes acting simultaneously are advection on the fast time scale $\tau = t$ and diffusion on the slow time scale $T = \epsilon t$. Thus we pose an asymptotic expansion

$$f(\theta, t; \epsilon) \quad \sim \quad f_0(\theta, \tau, T) + \epsilon f_1(\theta, \tau, T)$$

to be asymptotic at $T = \mathrm{ord}(1)$. Substituting into the governing equation and comparing coefficients of ϵ^n, we obtain in the usual way a sequence of problems.

At ϵ^0:

$$\frac{\partial f_0}{\partial \tau} + \frac{\partial}{\partial \theta}(\omega f_0) = 0$$

i.e.

$$\frac{1}{\omega}\left(\frac{\partial}{\partial \tau} + \omega\frac{\partial}{\partial \theta}\right)(\omega f_0) = 0$$

This equation says that an observer moving at a speed ω will see the quantity ωf_0 remain constant (on the fast time scale, only). It is therefore necessary to find out where an observer moving at ω has progressed to after a time τ. Let $\Theta(t)$ be the solution of the initial value problem

$$\dot{\theta} = \omega(\theta) \quad \text{in } t \geq 0 \text{ with } \theta = 0 \text{ at } t = 0$$

Because $\omega(\theta)$ is positive and 2π-periodic, $\Theta(t)$ modulo 2π must be periodic, say with period P.

We now transform from the co-ordinates θ and τ to the Lagrangian co-ordinates s and τ with

$$\theta(s,\tau) = \Theta(\tau - s)$$

The variable s is the time delay since θ was zero, and this is more useful than the usual Lagrangian variable of the initial angle. The variable is a periodic variable, with period P. Restricted to this period, there is an inverse

$$s = S(\theta, \tau)$$

To recast the differential equation in terms of the new co-ordinates s and τ, we first note

$$\delta\theta = \omega(\theta)(\delta\tau - \delta s)$$

so that

$$\left(\frac{\partial \theta}{\partial \tau}\right)_s = \omega(\theta) \quad \text{and} \quad \left(\frac{\partial s}{\partial \theta}\right)_\tau = -\frac{1}{\omega(\theta)}$$

Thence

$$\left(\frac{\partial}{\partial \tau}\right)_s = \left(\frac{\partial \tau}{\partial \tau}\right)_s\left(\frac{\partial}{\partial \tau}\right)_\theta + \left(\frac{\partial \theta}{\partial \tau}\right)_s\left(\frac{\partial}{\partial \theta}\right)_\tau = \left(\frac{\partial}{\partial \tau}\right)_\theta + \omega\left(\frac{\partial}{\partial \theta}\right)_\tau$$

$$\left(\frac{\partial}{\partial \theta}\right)_\tau = \left(\frac{\partial \tau}{\partial \theta}\right)_\tau\left(\frac{\partial}{\partial \tau}\right)_s + \left(\frac{\partial s}{\partial \theta}\right)_\tau\left(\frac{\partial}{\partial s}\right)_\tau = -\frac{1}{\omega}\left(\frac{\partial}{\partial s}\right)_\tau$$

Thus the controlling differential equation at ϵ^0 becomes

$$\left(\frac{\partial}{\partial \tau}\right)_s(\omega f_0) = 0$$

with solution

$$f_0(\theta, \tau, T) \;=\; \frac{A_0(s, T)}{\omega(\theta)}$$

This expresses formally the idea that on the fast time scale ωf_0 remains constant, moving with an observer at velocity ω, the observer being labelled by his value of s. This constant is, however, allowed to drift on the slow time scale. The initial value of the 'constant' is given from the initial conditions

$$A_0(s, 0) \;=\; F\left(\Theta(-s)\right) \omega\left(\Theta(-s)\right)$$

At ϵ^1:

$$\frac{\partial f_0}{\partial T} \;+\; \frac{\partial f_1}{\partial \tau} \;+\; \frac{\partial}{\partial \theta}\left(\omega f_1\right) \;=\; \frac{\partial^2 f_0}{\partial \theta^2}$$

In the transformed co-ordinates, and with the partially determined f_0, this becomes

$$\frac{1}{\omega}\frac{\partial A_0}{\partial T} \;+\; \frac{1}{\omega}\left(\frac{\partial}{\partial \tau}\right)_s\left(\omega f_1\right) \;=\; \frac{1}{\omega}\left(\frac{\partial}{\partial s}\right)_\tau \frac{1}{\omega}\left(\frac{\partial}{\partial s}\right)_\tau \frac{A_0}{\omega}$$

with now $\omega = \omega(\Theta(\tau - s))$, a P-periodic function of both s and τ. Thus

$$\frac{\partial A_0}{\partial T} \;+\; \left(\frac{\partial}{\partial \tau}\right)_s\left(\omega f_1\right) \;=$$

$$\frac{1}{\omega^2}\cdot\frac{\partial^2 A_0}{\partial s^2} \;+\; \frac{3}{2}\frac{\partial}{\partial s}\left(\frac{1}{\omega^2}\right)\cdot\frac{\partial A_0}{\partial s} \;+\; \frac{1}{2}\frac{\partial^2}{\partial s^2}\left(\frac{1}{\omega^2}\right)\cdot A_0$$

Now the right hand side of the equation is P-periodic in τ with a non-zero average. In order to maintain the asymptoticness of the expansion of f to $T = \mathrm{ord}(1)$, we must keep ωf_1 bounded as τ increases. Thus the average with respect to τ of the right hand side must be removed by setting it equal to $\partial A_0/\partial T$. Note that the second and third terms on the right hand side have zero averages, because they are derivatives with respect to s of functions of ω and so derivatives with respect to τ by $\omega = \omega(\Theta(\tau - s))$. Hence

$$\frac{\partial A_0}{\partial T} \;=\; \frac{1}{P}\int_0^P \frac{d\tau}{\omega^2\left(\Theta(\tau - s)\right)}\frac{\partial^2 A_0}{\partial s^2}$$

Our result is that the amplitude function $A_0(s, T)$ satisfies a simple diffusion equation with an effective diffusivity $< 1/\omega^2 >$. As this diffusivity is constant, the equation for A_0 can be readily solved for any particular initial condition F using Fourier Series over the P-periodic variable s.

The result $< 1/\omega^2 >$ for the effective diffusivity can be explained as follows. At lowest order f is conserved moving with speed ω. Thus where ω slows down, adjacent moving points crowd together with separations proportional to ω. This leads to the density of the conserved f increasing like $1/\omega$ as seen in the result for f_0. Now the steepening of the spatial gradients like $1/\omega$ enhances the diffusion like $1/\omega^2$. This enhanced diffusion acts ϵ slowly while f is being advected rapidly around the θ space. We thus need an average of the enhanced diffusivity, weighting with the time spent at each location.

7.4   Homogenised media

So far the differential equation has generated the two scales of interest. In this and the following sections the two scales are specified by the geometry. In this section we are concerned with the effective properties of a medium with some fine scale structure, e.g. the effective elastic moduli of a composite material with carbon fibre strengthening. Rather than the vector problem for the elastic displacement, we look at the simpler scalar problem of heat conduction,

$$\nabla \cdot k \cdot \nabla T = Q$$

with k and Q given functions which have a fine scale structure described by the short scale variable $\xi = x/\epsilon$. We pose an expansion which is to be asymptotic on the long scale $x = \text{ord}(1)$:

$$T(x, \epsilon) \sim T_0(\xi, x) + \epsilon T_1(\xi, x) + \epsilon^2 T_2(\xi, x)$$

Substituting into the governing equation and comparing coefficients of ϵ^n produces a sequence of problems.

At ϵ^{-2}:

$$\frac{\partial}{\partial \xi} \cdot k \cdot \frac{\partial T_0}{\partial \xi} = 0$$

with a solution

$$T_0 = T_0(x)$$

Thus at leading order, the temperature does not vary on the microscale.

At ϵ^{-1}:

$$\frac{\partial}{\partial \xi} \cdot k \cdot \frac{\partial T_1}{\partial \xi} = - \frac{\partial}{\partial \xi} \cdot k \cdot \frac{\partial T_0}{\partial x}$$

Thus T_1 will be linear in the forcing which is proportional to $\partial T_0/\partial x$, with a coefficient of linearity which depends on the details of $k(\xi)$, i.e.

$$T_1(\xi, x) \;=\; A(\xi) \cdot \frac{\partial T_0}{\partial x}$$

At ϵ^0:

$$\frac{\partial}{\partial \xi} \cdot k \cdot \frac{\partial T_2}{\partial \xi} \;=\; Q(\xi, x) - \frac{\partial}{\partial x} \cdot k \cdot \frac{\partial T_0}{\partial x} - \frac{\partial}{\partial \xi} \cdot k \cdot \frac{\partial T_1}{\partial x} - \frac{\partial}{\partial x} \cdot k \cdot \frac{\partial T_1}{\partial \xi}$$

To ensure that the expansion is asymptotic on $x = \mathrm{ord}(1)$, it is necessary to insist that the right hand side of the equation has a zero average over the fine scale details, otherwise T_2 would grow $\mathrm{ord}(\xi^2)$. Thus the secularity condition is

$$\frac{\partial}{\partial x} \cdot k^* \cdot \frac{\partial T_0}{\partial x} \;=\; Q^*$$

with

$$k^* \;=\; \left\langle k(\xi) + k(\xi) \cdot \frac{\partial A}{\partial \xi} \right\rangle \qquad \text{and} \qquad Q^* \;=\; \langle Q(\xi, x) \rangle$$

where the angled brackets denote an average over the ξ-microscale.

Note that our asymptotic analysis has shown that at leading order the temperature is constant over the local microstructure, and that to leading order the temperature satisfies a standard heat conduction equation with an effective heat conductivity k^* and heat source strength Q^*. Higher order corrections will find that there is a correction to the heat flux which is non-local, i.e. the heat flux at one point depends on the value of the temperature gradient in the neighbourhood of that point.

7.5   The WKBJ approximation

While everyone agrees that Messrs W, K, B and J did not invent this method, there is little agreement over who did. Certainly the following were all involved with important developments: Liouville 1837, Green 1837, Horn 1899, Rayleigh 1912, Gans 1915, Jeffrey 1923, Wentzel 1926, Kramers 1926, Brillouin 1926, Langer 1931, Olver 1961 and Meyer 1973. The problem is to obtain an asymptotic solution to the equation

$$\ddot{x} + f(\epsilon t)x \;=\; 0$$

We will tackle the problem with the method of multiple scales. Note that a more general equation

$$\ddot{y} + \epsilon a(\epsilon t)\dot{y} + b(\epsilon t)y \;=\; 0$$

can be transformed to our canonical form by the substitution

$$y = x \exp\left(\tfrac{1}{2}\epsilon \int^t a(\epsilon t')\, dt'\right)$$

resulting in

$$\ddot{x} + \left(b - \tfrac{1}{2}\epsilon^2 a'' - \tfrac{1}{4}\epsilon^2 a^2\right) x = 0$$

7.5.1   Solution by multiple scales

First we consider the case of $f > 0$, so that we may put $f = \omega^2$ with ω real and positive. Then the obvious solution is a fast oscillation with a local frequency ω and a slowly drifting amplitude and phase. In the notation of the method of multiple scales, with fast time $\tau = t$ and slow time $T = \epsilon t$, we expect a solution at leading order of the form

$$R(T) \cos\left[\omega(T)\tau + \theta(T)\right]$$

Unfortunately the secularity condition, which is found in the problem for x_1

$$\theta_T = -\omega_T \tau$$

is unacceptable because of the occurrence of the fast variable in the drift equation, with the other factors depending only on the slow time. The failure of our attempted solution does, however, suggest a cure. If the phase θ had been much larger, $\mathrm{ord}(\epsilon^{-1})$ rather than $\mathrm{ord}(1)$, then we could have multiplied the above equation by ϵ to produce an acceptable equation, which only involved the slow time T.

We thus start again with a leading order solution of the form

$$x_0(\tau, T) = R(T) \cos\theta \quad\text{with}\quad \theta = \epsilon^{-1}\Theta_0(T) + \Theta_1(T)$$

It is not immediately apparent that this solution does oscillate on the fast time scale. Note however that the time derivative of the phase is

$$\theta_t = \Theta_{0T} + \epsilon\Theta_{1T}$$

which is order 1. Thus for an order 1 change in time t, there is a small relative change in the slowly varying $\Theta_0(T)$, but this small relative change of a large object results in an order 1 absolute change, and for the arguments of trigonometric functions it is the magnitude of the absolute change which is relevant. It was this consideration which also required the third unknown function Θ_1 term to be included in the leading order term for $x(t, \epsilon)$.

Evaluating the time derivatives of the above x_0 we have

$$\dot{x}_0 = -R\Theta_{0T}\sin\theta + \epsilon\left[R_T\cos\theta - R\Theta_{1T}\sin\theta\right]$$
$$\ddot{x}_0 = -R\Theta_{0T}^2\cos\theta + \epsilon\left[-(2R_T\Theta_{0T} + R\Theta_{0TT})\sin\theta\right.$$
$$\left.- 2R\Theta_{0T}\Theta_{1T}\cos\theta\right] + O(\epsilon^2)$$

Substituting into the governing equation yields at leading order

$$\Theta_{0T} = \omega$$

At the next order, the secularity conditions associated with the equation for x_1 are

$$2R_T\Theta_{0T} + R\Theta_{0TT} = 0$$
$$2R\Theta_{0T}\Theta_{1T} = 0$$

with solutions

$$R^2\omega = \text{constant}$$
$$\Theta_1 = \text{constant}$$

Note that it is not the energy $E = \frac{1}{2}R^2\omega^2$ which is conserved over the long time scales, but rather the action E/ω (sometimes called an adiabatic invariant).

Note that once the secularity conditions are satisfied, then there is no forcing in the equation for x_1, although there is some forcing for x_2. Thus if one were to satisfy the boundary conditions with x_0 to order ϵ instead of just to order 1, then the error of x_0 would be only $O(\epsilon^2)$.

Returning to the original canonical equation, we now have a solution when $f > 0$ for x which is asymptotic to $t = \text{ord}(\epsilon^{-1})$. This solution is often expressed as

$$x(t,\epsilon) \sim [f(\epsilon t)]^{-1/4}\,(a\cos\theta + b\sin\theta) \quad \text{with } \theta = \int_0^t [f(\epsilon t')]^{1/2}\,dt'$$

with constants a and b. Similarly when $f < 0$, the solution takes the form

$$x(t,\epsilon) \sim [-f(\epsilon t)]^{-1/4}\,\left(Ae^\varphi + Be^{-\varphi}\right) \quad \text{with } \varphi = \int_0^t [-f(\epsilon t')]^{1/2}\,dt'$$

with constants A and B.

7.5.2   Higher approximations

To obtain higher approximations it would be necessary to introduce a hierarchy of super slow time scales $T_n = \epsilon^n t$. Here we avoid these and instead use a method particular to the WKBJ problem. The first order asymptotic theory suggests a transformation

$$x(t,\epsilon) \;\equiv\; \mathrm{Re}\left\{r(\epsilon t, \epsilon)\exp\left[i\int^t \sigma(\epsilon t', \epsilon)\,dt'\right]\right\}$$

with r and σ required to be real quantities. Then dropping the real part sign

$$\dot{x} \;=\; ir\sigma\exp + \epsilon r_T \exp$$
$$\ddot{x} \;=\; -r\sigma^2\exp + \epsilon i(2r_T\sigma + r\sigma_T)\exp + \epsilon^2 r_{TT}\exp$$

Substituting into the governing equation and comparing real and imaginary parts, we find

$$2r_T\sigma + r\sigma_T \;=\; 0$$
$$\epsilon^2 r_{TT} + r\left(f - \sigma^2\right) \;=\; 0$$

The first equation can be integrated to give the general result

$$r^2\sigma \;=\; \text{constant}$$

The second equation is then a nonlinear differential equation for the amplitude r. To solve it we expand in powers of ϵ^2:

$$\sigma(T,\epsilon) \;\sim\; \sigma_0(T) \;+\; \epsilon^2\sigma_1(T)$$
$$r(T,\epsilon) \;\sim\; r_0(T) \;+\; \epsilon^2 r_1(T)$$

Substituting into the nonlinear differential equation and comparing coefficients of ϵ^n, we obtain

$$\sigma_0 \;=\; f^{1/2}$$
$$r_0 \;=\; kf^{-1/4} \quad \text{with } k \text{ a constant}$$
$$\sigma_1 \;=\; \frac{r_{0TT}}{2\sigma_0 r_0} \;=\; f^{1/2}\left[\frac{5f'^2}{32f^3} - \frac{f''}{8f^2}\right]$$
$$r_1 \;=\; -\frac{r_0\sigma_1}{2\sigma_0}$$

Exercise 7.7.   Use the transformation at the beginning of §7.5.2 to obtain solutions of the WKBJ type to the fourth order equation

$$\ddddot{x} + f(\epsilon t)x \;=\; 0$$

7.5.3   Turning points

Our solutions in §7.5.1 work well while $f > 0$ or $f < 0$, but there is trouble at $f = 0$ where the frequency vanishes and the amplitude becomes infinite. Without loss of generality we can move the point where f vanishes, the so-called turning point, to the origin. Thus we have $f(0) = 0$. We start with the case $f'(0) > 0$, and leave other cases to the end of this section.

Now far from the origin, where $\epsilon t = \text{ord}(1)$, we have the solutions given at the end of §7.5.1; the trigonometric solutions being applicable in $t > 0$, and the exponential solutions in $t < 0$. The aim of this section is to produce a solution which is valid near $t = 0$, so that we can provide a connection between the constants a and b in the trigonometric region to the constants A and B in the exponential region. We are thus involved in a problem of matched asymptotic expansions.

Now near to the origin, where $|\epsilon t| \ll 1$, we can approximate the governing equation by

$$\ddot{x} + \epsilon t f'(0)x = 0$$

If we introduce the rescaling

$$\tau = -t \left(\epsilon f'(0)\right)^{1/3}$$

we recover Airy's equation

$$x_{\tau\tau} - \tau x = 0$$

with a general solution

$$x = \alpha \text{Ai}(\tau) + \beta \text{Bi}(\tau)$$

in which α and β are constants and Ai and Bi are Airy functions. The asymptotic behaviour of Ai at large (positive and negative) arguments was evaluated in §3.3 by the method of steepest descents. The second Airy function Bi can be treated similarly.

Matching for negative times as $\tau \nearrow +\infty$ in the inner and as $t \nearrow 0$ in the outer

$$\text{inner} = \frac{1}{\tau^{1/4}\sqrt{\pi}} \left(\tfrac{1}{2}\alpha \exp(-\tfrac{2}{3}\tau^{3/2}) + \beta \exp(\tfrac{2}{3}\tau^{3/2})\right)$$

$$\text{outer} = \frac{1}{[-\epsilon t f'(0)]^{1/4}} \left(A \exp(\varphi) + B \exp(-\varphi)\right)$$

$$\text{where } \varphi = -\tfrac{2}{3}\left[\epsilon f'(0)\right]^{1/2} (-t)^{3/2}$$

The matching is successful if we take for the constants

$$\alpha = \frac{2\sqrt{\pi}}{[\epsilon f'(0)]^{1/6}} A \quad \text{and} \quad \beta = \frac{\sqrt{\pi}}{[\epsilon f'(0)]^{1/6}} B$$

Now matching for positive times as $\tau \searrow -\infty$ in the inner and as $t \searrow 0$ in the outer

$$\text{inner} = \frac{1}{(-\tau)^{1/4}\sqrt{\pi}} (\alpha \sin \Theta + \beta \cos \Theta)$$

$$\text{where } \Theta = \tfrac{2}{3}(-\tau)^{3/2} + \tfrac{1}{4}\pi$$

$$\text{outer} = \frac{1}{[\epsilon t f'(0)]^{1/4}} (a \cos \theta + b \sin \theta)$$

$$\text{where } \theta = \tfrac{2}{3}[\epsilon f'(0)]^{1/2} t^{3/2}$$

The matching is again successful if we take for the constants

$$a = \frac{[\epsilon f'(0)]^{1/6}}{(2\pi)^{1/2}} (\alpha + \beta) \quad \text{and} \quad b = \frac{[\epsilon f'(0)]^{1/6}}{(2\pi)^{1/2}} (\alpha - \beta)$$

Thus we have obtained the important *connection formulae*

$$A = \frac{a+b}{2\sqrt{2}} \quad \text{and} \quad B = \frac{a-b}{\sqrt{2}}$$

So if x is exponentially small in $t < 0$, i.e. $B = 0$ (because $\varphi < 0$), we emerge into $t > 0$ with $a \sim b$, i.e. with a solution

$$2Af^{-1/4} \sin\left(\int_0^t f^{1/2} + \frac{\pi}{4}\right)$$

showing a phase shift of $\pi/4$ for coming through the turning point.

We have studied above the case with $f'(0) > 0$. The case with $f'(0) < 0$ just requires a reversal of the t co-ordinate. For higher order turning points with $f \sim k^2 t^n$ as $t \to 0$, one has a governing equation in the inner region

$$\ddot{x} + k^2 t^n x = 0$$

which has solutions in terms of Bessel functions $t^{1/2} J_{\pm\nu}(2k\nu t^{1/2\nu})$ where $\nu = 1/(2+n)$.

If the details of the solution in the neighbourhood of the turning point are not required, there is an alternative way to derive the connection formulae. It is possible to go from the region where $f < 0$ to the region where $f > 0$ avoiding the point where $f = 0$ by making a detour on the complex t-plane. Caution is needed because of a Stokes phenomenon in which different asymptotic expansions are restricted to different sectors in the complex t-plane. The origin of the $\pi/4$ phase shift is however seen as the analytic continuation of $(-f)^{-1/4}e^{\varphi}$ to $(f)^{-1/4}e^{-i(\theta-\pi/4)}$.

7.5.4 Examples

Example 1 is to find the high energy eigensolutions of Schrödinger's equation for a simple harmonic oscillator. The problem is to solve

$$\psi'' + (E - x^2)\psi = 0$$

with $\psi \to 0$ as $x \to \pm\infty$

when the eigenvalue E is large.

Now when E is large, there will be an oscillatory solution in $x^2 < E$ with a wavelength $\mathrm{ord}(E^{-1/2})$ which is short compared with the scale on which this wavelength changes, $\mathrm{ord}(E^{1/2})$. Thus we have a WKBJ problem. There are turning points at $x = \pm E^{1/2}$, and we want the exponentially decaying solution beyond these turning points.

Then by our WKBJ theory we have in $x^2 < E$

$$\psi \sim \frac{1}{(E - x^2)^{1/4}} \sin\left(\int_{-\sqrt{E}}^{x} (E - x^2)^{1/2}\,dx + \frac{\pi}{4}\right)$$

using the turning point connection formula for the decaying solution in $x < -\sqrt{E}$. Requiring the solution to decay also in $x > \sqrt{E}$ via a connection formula at $x = \sqrt{E}$, we have

$$\int_{-\sqrt{E}}^{\sqrt{E}} (E - x^2)^{1/2}\,dx + \frac{\pi}{4} = n\pi - \frac{\pi}{4}$$

the minus sign coming from the reversing of the x-coordinate at the second turning point. The above equation determines the eigenvalues to be

$$E = 2n - 1$$

which happens to be exact for all n, rather than just asymptotically true for large E. Our WKBJ solution also easily gives the behaviour

$$\psi \sim x^{n-1}e^{-x^2/2} \quad \text{as } x \to \pm\infty$$

as well as the behaviour in $x^2 < E$.

Example 2 is to find the large eigenvalue solutions of the Legendre's equation. The standard form of Legendre's equation (to be solved for the solution which is regular at $x = -1$ and 1)

$$[(1 - x^2)\, y']' + \lambda y = 0$$

can be transformed into our canonical WKBJ form by setting

$$y = Y \exp\left(\int \frac{x}{(1 - x^2)} \right) = Y(1 - x^2)^{-1/2}$$

to give

$$Y'' + \left[\frac{\lambda}{1 - x^2} + \frac{1}{(1 - x^2)^2} \right] Y = 0$$

As in the first example, the solution has a short scale oscillation when λ is large. Thus in the oscillating range $-1 < x < 1$ our WKBJ solution gives

$$Y \sim k(1 - x^2)^{1/4} \sin\left[\lambda^{1/2} \int_{-1}^{x} (1 - x^2)^{-1/2}\, dx + \theta \right]$$

with the phase θ to be found from a turning point analysis.

Note that the wavenumber $[\lambda/(1 - x^2)]^{1/2}$ is singular at the ends $x = \pm 1$, although in an integrable way. This singular behaviour, plus the more singular correction $(1 - x^2)^{-2}$ in the wavenumber squared, calls for a modified turning point analysis. The regular solution with $y(-1) = 1$ has a solution near $x = -1$

$$Y \sim \sqrt{2}(1 + x)^{1/2} J_0\left([2\lambda(1 + x)]^{1/2} \right)$$

Matching this to our solution away from the end $x = -1$, we find the ubiquitous phase shift $\theta = \pi/4$ and also the constant $k = 2(2\pi\lambda)^{-1/4}$. Applying a similar analysis at $x = 1$, we obtain the eigenvalue condition

$$\lambda^{1/2}\pi + \tfrac{1}{4}\pi = (n + 1)\pi - \tfrac{1}{4}\pi$$

$$\text{i.e. } \lambda = n(n + 1) + \tfrac{1}{4}$$

i.e differing from the exact result through the error of $1/4$.

Exercise 7.8.   Find the large eigenvalue solutions of the equation

$$y'' + \lambda(1 - x^2)^2 y = 0$$

subject to $y = 0$ at $x = \pm 1$.

At the ends $x = \pm 1$ you will need to use turning point solutions like $(1-x^2)^{1/2}J_{1/4}(\lambda^{1/2}(1-x^2)^2/4)$, and then use $J_{1/4}(z) \sim (2/\pi z)^{1/2}\cos(z - 3\pi/8)$ as $z \to \infty$.

7.5.5 Use of the WKBJ approximation to study an exponentially small term

Consider the matched asymptotic expansion problem for $y(x; \epsilon)$

$$\epsilon y'' - a(x)y' + b(x)y = 0 \quad \text{in} \quad -1 \le x \le 1$$

$$\text{with} \quad y(-1) = A \quad \text{and} \quad y(1) = B$$

with the function $a(x)$ such that $a > 0$ near $x = 1$ and $a < 0$ near $x = -1$. This restriction on the behaviour of a is crucial to the structure of this problem. To simplify the analysis we require additionally that $a > 0$ in $x > 0$ and $a < 0$ in $x < 0$, and that at $a(0+) = -a(0-)$ and $b(0+) = -b(0-)$.

Simple use of the method of matched asymptotic expansions produces an outer solution for the interior $-1 < x < 1$

$$y \sim k \exp \int_0^x \frac{b(x)}{a(x)} \, dx$$

with corrections ord(ϵ). Note that the additional restrictions on a and b make y and y' continuous at $x = 0$. In order to satisfy the two boundary conditions, inner solutions are required, and these are possible near both boundaries because of the restriction on the sign of a. The inner near $x = -1$ is

$$y \sim A + \left(k \exp \int_0^{-1} \frac{b}{a} \, dx - A \right) \left[1 - \exp \frac{a(-1)(1+x)}{\epsilon} \right]$$

while the inner near $x = 1$ is

$$y \sim B + \left(k \exp \int_0^1 \frac{b}{a} \, dx - B \right) \left[1 - \exp \frac{a(1)(1-x)}{\epsilon} \right]$$

There now appears to be a paradox: the above matched asymptotic solution appears to be valid for all values of k, whereas the original equation had a unique solution. Proceeding to higher order corrections does not help to determine k.

The resolution of the paradox comes from realising that the second, rapidly decaying solution of the differential equation near to $x = -1$ is related to the second, rapidly decaying solution near to $x = 1$. One is therefore not at liberty to pick the amplitude as $(k \exp -A)$ at one end and as $(k \exp -B)$ at the other end. To find out how the amplitude of this second, rapidly decaying solution is related from one end to the other, we use the WKBJ approximation.

The rapidly varying (decaying) second solution of the differential equation is for $x \neq 0$

$$y \quad \sim \quad \left(\exp \int_0^x \frac{a(x')}{\epsilon} \, dx' \right) \frac{1}{|a|} \left(\exp - \int_0^x \frac{b(x')}{a(x')} \, dx' \right)$$

This needs modifying in the neighbourhood of $x = 0$ in order to make y' continuous there.

We can now conclude that there is an inner solution only near $x = -1$ or only near $x = 1$ according to whether

$$\int_{-1}^1 a(x') \, dx' \quad \lessgtr \quad 0$$

7.5.6   The small reflected wave
in the WKBJ approximation

Consider waves propagating in one dimension through a medium whose properties vary slowly with position, i.e.

$$y_{tt} \;=\; \left(c(\epsilon x)^2 y_x \right)_x$$

Then the WKBJ solutions are

$$y \sim A c^{-1/2} e^{it - i\theta} \;+\; B c^{-1/2} e^{it + i\theta}$$

$$\text{where} \quad \theta \;=\; \int_0^x \frac{dx'}{c(\epsilon x')}$$

The A-term and B-term represent waves travelling respectively to the right and to the left. The variation of the amplitude like $c^{-1/2}$ means that the flux of energy for each wave, $c^2 y_x y_t$, is constant. Thus when a single right-moving wave is incident on a region where the medium varies, $c(\epsilon x)$, all of its energy is transmitted and there is no reflected wave, to leading order.

Now if c has a discontinuity (where it is not slowly varying), then equating y and $c^2 y_x$ on the two sides of the discontinuity yields a reflected wave with a relative magnitude $\mathrm{ord}(\Delta c / c)$.

Now if c is continuous but c_x is discontinuous, then equating y and $c^2 y_x$ on the two sides of the discontinuity yields a smaller reflected wave $\mathrm{ord}(\epsilon \Delta c')$.

Using the higher order solutions of §7.5.2, one can show that if $c(\epsilon x)$ has a discontinuity in its nth derivative, then there will be a reflected wave of order ϵ^n.

If all the derivatives of $c(\epsilon x)$ are continuous (a C^∞ function), then the reflected wave is exponentially small. This exponentially small reflected wave can be calculated by making a transformation

$$y(x,t;\epsilon) \equiv A(x)c^{-1/2}e^{it-i\theta} + B(x)c^{-1/2}e^{it+i\theta}$$

Following the method of variations of parameters, this transformation yields equations for A and B, the energy amplitudes of the transmitted and reflected waves,

$$A_x = i\epsilon^2 g\left(A + Be^{2i\theta}\right)$$
$$B_x = -i\epsilon^2 g\left(B + Ae^{-2i\theta}\right)$$
$$\text{with } g = (c'' + c'^2/2c)/4$$

A further transformation $A = a(x)e^{i\epsilon^2\varphi}$ and $B = b(x)e^{-i\epsilon^2\varphi}$ with $\varphi = \int_0^x g(\epsilon x')\,dx'$ yields

$$b_x = -i\epsilon^2 gae^{-2i\theta+i2\epsilon^2\varphi}$$

In this equation one may take a to be a constant when the reflected wave b is small ($b \ll a$). The integral for the change in b can then be evaluated by deforming the contour on the complex x-plane. The major contribution will come from the complex singularity of $c(\epsilon x)$ at $\epsilon x = X_*$ which has the smallest real part of

$$I(X_*) = \frac{1}{i}\int_0^{X_*} \frac{dX}{c(X)}$$

If the nearest singularity is a pole of c, the reflected wave is found to be

$$b(-\infty) = -i\epsilon ae^{-2I(X_*)/\epsilon}\left[1.41 + O(\epsilon^{1/2})\right]$$

7.6   Slowly varying waves

7.6.1   A model problem

We consider a model problem due to Bretherton which describes waves propagating with some dispersion, with a small nonlinearity, and in a slowly varying medium.

$$\varphi_{tt} + (\alpha\varphi_{xx})_{xx} + (\beta\varphi_x)_x + \gamma\varphi = \epsilon\varphi^3$$

where α, β and γ are given functions of a slow space variable $X = \epsilon x$ and a slow time $T = \epsilon t$. The solution will clearly be a wave with a local frequency and a local wavenumber, and with a slowly varying amplitude. Now the basic oscillation of the wave has an order 1 change in the basic phase. In order to produce this order 1 change and also to have derivatives which vary only slowly in space and time (the slowly varying wave number and frequency), we rescale the phase to be very large and allow this large phase to have only slow variations. Thus we pose

$$\varphi(x,t;\epsilon) \sim a(X,T)\cos\theta \quad \text{with} \quad \theta = \epsilon^{-1}\Theta_0(X,T) + \Theta_1(X,T)$$

The *local wavenumber* is then $k(X,T) \equiv \theta_x \sim \Theta_{0X}$ and the *local frequency* is $\omega(X,T) \equiv -\theta_t \sim -\Theta_{0t}$. Note that this is the appropriate generalisation of the integral $\int^t \omega(\epsilon t')\,dt'$ in the WKBJ method.

With the assumed asymptotic form of the solution for φ, various time and space derivatives can be evaluated, e.g.

$$\varphi_{0xxxx} = ak^4\cos\theta + \epsilon\big[(4a_X k^3 + 6ak^2 k_X)\sin\theta$$
$$+ 4ak^3\Theta_{1X}\cos\theta\big] + O(\epsilon^2)$$

Substituting these into the governing equation yields at ϵ^0 just the *dispersion relation*

$$\omega^2 = \alpha k^4 - \beta k^2 + \gamma$$

At next order, the need to avoid a resonating forcing like $\sin\theta$ requires

$$(2a_T\omega + a\omega_T) + \alpha(4a_X k^3 + 6ak^2 k_X) + 2\alpha_X ak^3$$
$$- \beta(2a_X k + ak_X) - \beta_X ak = 0$$

And avoiding a resonating forcing like $\cos\theta$ requires

$$2a\omega\Theta_{1T} + \alpha 4ak^3\Theta_{1X} - \beta 2ak\Theta_X = \tfrac{3}{4}a^3$$

We have at this stage to assume that the third harmonic $\cos 3(kx - \omega t)$ is not also resonant, which it can be for some dispersion relations – see the exercise below.

To make some sense out of our secularity conditions, we introduce the *group velocity* $c = \partial\omega/\partial k$. Differentiating the dispersion relation with respect to k, we have

$$2\omega c = 4\alpha k^3 - 2\beta k$$

We can now recognise the $\cos\theta$ secularity condition as

$$\left(\frac{\partial}{\partial T} + c\frac{\partial}{\partial X}\right)\Theta_1 = \frac{3a^2}{8\omega}$$

i.e. the extra phase Θ_1 drifts as in Duffing's equation, as seen by an observer moving with the group velocity.

Multiplying the $\sin \theta$ secularity condition by a and regrouping, we have

$$\frac{\partial}{\partial T}\left(a^2\omega\right) + \frac{\partial}{\partial X}\left(ca^2\omega\right) = 0$$

i.e. moving with the group velocity the quantity $a^2\omega$ is conserved. Note that this quantity is (energy $= a^2\omega^2)/\omega$ and is known as *wave action*.

We now have relations for the drift in the amplitude and extra phase. To complete the description we need how the wave number and frequency vary as they obey the dispersion relation. If we differentiate the dispersion relation with respect to time, we find

$$2\omega\omega_T = (4\alpha k^3 - 2\beta k)k_T + \alpha_T k^4 - \beta_T k^2 + \gamma_T$$

Now $k \equiv \Theta_X$, so that $k_T = \Theta_{XT} = -\omega_X$. Thus we have

$$\left(\frac{\partial}{\partial T} + c\frac{\partial}{\partial X}\right)\omega = (\alpha_T k^4 - \beta_T k^2 + \gamma_T)/2\omega$$

i.e. moving with the group velocity the frequency changes due to slow time changes in the medium. Similarly differentiating the dispersion relation with respect to space and using again $\omega_X \equiv -k_T$, we find

$$\left(\frac{\partial}{\partial T} + c\frac{\partial}{\partial X}\right)k = -(\alpha_X k^4 - \beta_X k^2 + \gamma_X)/2\omega$$

i.e. moving with the group velocity the wavenumber changes due to slow spatial variations in the medium.

Our model can be generalised to a moving medium with a slowly varying velocity $U(X,T)$, by replacing the partial double derivative with respect to time ∂_t^2 with the self-adjoint advected double derivative $(\partial_t + \partial_x U)(\partial_t + U\partial_x)$. Thus in the governing equation the term φ_{tt} becomes

$$\varphi_{tt} + 2U\varphi_{xt} + U^2\varphi_{xx} + \epsilon[U_X\varphi_t + (U_T + 2UU_X)\varphi_x] + O(\epsilon^2)$$

The effect of this modification on the dispersion relation is to turn ω^2 into $\omega^2 - 2U\omega k + U^2 k^2$, i.e. $(\omega - Uk)^2$. Hence we find in the moving medium

$$\omega = Uk + \omega^+(k)$$

where ω^+ is the intrinsic frequency of the stationary medium. Thus the group velocity becomes $c = U + c^+$, where c^+ is the intrinsic group velocity of the stationary medium. The moving medium changes the term $2a_T\omega + a\omega_T$ in the $\sin\theta$ secularity condition to

$$[2a_T\omega + a\omega_T] + 2U\left[-a_T k + a_X\omega + \tfrac{9}{2}(\omega_X - k_T)\right]$$

$$+ U^2 \left[-2a_X k - ak_X \right] + U_X a\omega - \left[U_T + 2UU_X \right] ak$$

$$= \frac{1}{a} \left[\left[a^2(\omega - Uk) \right]_T + \left(a^2(\omega - Uk) \right)_X \right]$$

Adding this to the remaining unmodified term $\left(a^2\omega^+ c^+ \right)_X / a$ yields

$$\frac{\partial}{\partial T} \left(a^2\omega^+ \right) + \frac{\partial}{\partial X} \left(ca^2\omega^+ \right) = 0$$

i.e. the wave action $a^2\omega^+$ is conserved moving with the group velocity. Similarly one can show that the drift equation for the extra phase Θ_1 is modified by replacing ω by ω^+ on the right hand side, and that the wave number and frequency drift equations are modified by a similar replacement of ω by ω^+ together with additional terms $-U_X k$ and $U_T k$ added to the right hand sides.

Exercise 7.9 on two waves resonantly interacting. Two possible solutions of the partial differential equations

$$5\psi_{tt} + \psi_{xxxx} + 4\psi = 0$$

are the waves $\cos(x - t)$ and $\cos(2x - 2t)$.
(i)   Obtain the first order partial differential equations which govern the slow drift in the amplitudes of these two waves on the space and time scales of order ϵ^{-1} for the weak interaction between wave packets governed by

$$5\psi_{tt} + \psi_{xxxx} + 4\psi = \epsilon\psi\psi_x$$

You may neglect the slower drift in the phases.
(ii)   Look for the steady periodic solutions of the equation

$$5\psi_{tt} + \psi_{xxxx} + 4\psi = \epsilon\psi^2$$

which take the form

$$A \cos \left[(1 + \epsilon k)x - t \right] + B \cos 2 \left[(1 + \epsilon k)x - t \right] + O(\epsilon)$$

obtaining relationships between A, B and k.

7.6.2   Ray theory

We now move from the particular model problem with its unpleasant algebraic details to a general theory of waves propagating through a slowly varying medium. The first part which describes how the local wavenumber and frequency change is variously called ray theory, wave kinematics or geometric optics.

We consider a scalar wave field $\varphi(\mathbf{x}, t)$ in three-dimensional space. We assume that asymptotically it takes the form of a single wave propagating with a slowly varying amplitude and slowly varying wavenumber and frequency,

$$\varphi(\mathbf{x}, t; \epsilon) \sim a(\mathbf{X}, T) \cos\theta \quad \text{with} \quad \theta = \epsilon^{-1}\Theta(\mathbf{X}, T)$$

This wave has a local frequency $\omega = -\theta_t = -\Theta_T$ and a local wavenumber $\mathbf{k} = \partial\theta/\partial\mathbf{x} = \partial\Theta/\partial\mathbf{X}$. Immediately from these definitions we have two consistency relations – known as the *conservation of wave crests* in space and time:

$$\frac{\partial}{\partial\mathbf{X}} \wedge \mathbf{k} = 0 \quad \text{and} \quad \frac{\partial\omega}{\partial\mathbf{X}} + \frac{\partial\mathbf{k}}{\partial T} = 0$$

Now the local dynamics will produce a dispersion relation

$$\omega = \Omega(\mathbf{k}, \alpha(\mathbf{X}, T))$$

where the dependence on the medium is shown by a dependence on the parameter α. (Note that α could also represent the amplitude of the wave in a nonlinear problem.) The group velocity is defined by $\mathbf{c} = \partial\Omega/\partial\mathbf{k}$. Differentiating the dispersion relation with respect to time, we find

$$\frac{\partial\omega}{\partial T} = \frac{\partial\Omega}{\partial\mathbf{k}}\frac{\partial\mathbf{k}}{\partial T} + \frac{\partial\Omega}{\partial\alpha}\frac{\partial\alpha}{\partial T}$$

Using the consistency relation for $\mathbf{k}_T$, this becomes

$$\frac{\partial\omega}{\partial T} + \mathbf{c} \cdot \frac{\partial\omega}{\partial\mathbf{X}} = \frac{\partial\Omega}{\partial\alpha}\frac{\partial\alpha}{\partial T}$$

Exercise 7.10. Derive the similar result

$$\frac{\partial\mathbf{k}}{\partial T} + \left(\mathbf{c} \cdot \frac{\partial}{\partial\mathbf{X}}\right)\mathbf{k} = -\frac{\partial\Omega}{\partial\alpha}\frac{\partial\alpha}{\partial\mathbf{X}}$$

7.6.3 Averaged Lagrangian

The second part of the general behaviour of waves propagating in a slowly varying medium concerns the conservation of wave action, which gives how the amplitude varies.

In a continuum, the Lagrangian L is expressed in terms of a Lagrangian density $\mathcal{L}$, so $L = \int \mathcal{L}\,dx\,dt$. In this subsection we work in just one space dimension, three dimensions being a trivial extension.

The Lagrangian density will depend on the scalar wave field φ and its derivatives, and parametrically on the slow space and time as the medium varies, i.e. $\mathcal{L} = \mathcal{L}(\varphi, \varphi_t, \varphi_x, \varphi_{xx}, \ldots; X, T)$. For linear waves $\mathcal{L}$ is quadratic in φ. In our model problem,

$$\mathcal{L} = \tfrac{1}{2}\left((\varphi_t + U\varphi_x)^2 - \alpha\varphi_{xx}^2 + \beta\varphi_x^2 - \gamma\varphi^2 + \tfrac{1}{2}\epsilon\varphi^4\right)$$

The field equations for the wave field φ follow as the Euler–Lagrange equations, corresponding to the vanishing of the first variation of L with respect to φ, i.e.

$$\frac{\partial\mathcal{L}}{\partial\varphi} - \frac{\partial}{\partial t}\left(\frac{\partial\mathcal{L}}{\partial\varphi_t}\right) - \frac{\partial}{\partial x}\left(\frac{\partial\mathcal{L}}{\partial\varphi_x}\right) + \frac{\partial^2}{\partial x^2}\left(\frac{\partial\mathcal{L}}{\partial\varphi_{xx}}\right) + \cdots = 0$$

We now substitute a solution of a single slowly varying wave

$$\varphi(x,t;\epsilon) \sim a(X,T)\cos\theta \quad \text{with} \quad \theta = \epsilon^{-1}\Theta(X,T)$$

with local wavenumber $k = \theta_x = \Theta_X$ and local frequency $\omega = -\theta_t = -\Theta_T$. Our Lagrangian density becomes $\mathcal{L}(a\cos\theta, a\omega\sin\theta, -ak\sin\theta, -ak^2\cos\theta, \ldots; X, T)$, at least asymptotically.

Now the full Lagrangian L is evaluated by integrating the density $\mathcal{L}$ over all x and t. In this integration there is a slow variation with X and T, and a fast variation in θ. The integration therefore averages over the rapid θ variations, producing an effective averaged Lagrangian density $\overline{\mathcal{L}}$

$$L = \epsilon^{-2}\int\overline{\mathcal{L}}\,dX\,dT \quad \text{with} \quad \overline{\mathcal{L}} = \frac{1}{2\pi}\int\mathcal{L}(a,\theta; X,T)\,d\theta$$

Thus in our model problem

$$\overline{\mathcal{L}} = \tfrac{1}{2}a^2\left((\omega - Uk)^2 - \alpha k^4 + \beta k^2 - \gamma - \tfrac{3}{16}\epsilon a^2\right)$$

The dynamics for the wave are derived from Lagrange's principle that the first variations of L with respect to the generalised co-ordinates a and Θ must vanish. (Note that Θ enters only through its derivatives ω and k.) The variation with respect to the amplitude a,

$$\frac{\partial\overline{\mathcal{L}}}{\partial a} = 0$$

is the dispersion relation. In our model problem this is

$$(\omega - Uk)^2 = \alpha k^4 - \beta k^2 + \gamma + \tfrac{3}{8}\epsilon a^2$$

In a linear wave theory $\overline{\mathcal{L}} = \tfrac{1}{2}a^2 F(\omega, k)$, with F the dispersion relation. Hence the first variation with respect to the amplitude a gives $F = 0$.

This therefore implies that there is an equipartion of energy between the average potential and average kinetic energies.

The first variation with respect to the phase Θ gives

$$-\frac{\partial}{\partial T}\left(-\frac{\partial \overline{\mathcal{L}}}{\partial \omega}\right) - \frac{\partial}{\partial X}\left(\frac{\partial \overline{\mathcal{L}}}{\partial k}\right) = 0$$

This is a conservation equation which says that the density $-\partial \overline{\mathcal{L}}/\partial \omega$ changes due to a divergence in the flux $\partial \overline{\mathcal{L}}/\partial k$. Now for linear waves

$$\frac{\partial \overline{\mathcal{L}}}{\partial k} = \tfrac{1}{2}a^2 \frac{\partial F}{\partial k} = -\tfrac{1}{2}a^2 \frac{\partial F}{\partial \omega}\left(\frac{\partial \omega}{\partial k}\right)_{F=0} = -\frac{\partial \overline{\mathcal{L}}}{\partial \omega} c$$

Thus the variation with respect to Θ for linear waves becomes

$$\frac{\partial A}{\partial T} + \frac{\partial}{\partial X}(cA) = 0$$

i.e. the wave action $A = \partial \overline{\mathcal{L}}/\partial \omega$ is conserved moving with the group velocity $c = \partial \omega/\partial k$ evaluated on the dispersion relation $F = 0$.

Improved convergence

In most new problems one only calculates the leading order approximation, particularly in a difficult asymptotic analysis involving matched
expansions or multiple scales. In some old important problems, several
terms have now been found. Recently in some regular problems, computers have been set to find a large number of terms, with the computers
performing algebra rather than arithmetic.

The hope is that several terms will give good quantitative results when
the small parameter is not so small. This is likely to be more true of
a convergent expansion than an asymptotic one, but even a convergent
series can converge slowly. We now look at the limits of convergence and
the possibilities of rearranging an expansion to improve its convergence.

8.1 Radius of convergence and the Domb–Sykes plot

We suppose that the result of an asymptotic study can be expressed in
a power series in a small positive parameter ϵ,

$$f(\epsilon) \quad \sim \quad \sum_{n=0}^{N} c_n \epsilon^n$$

Thus we exclude from consideration those studies which have both ϵ^α
and $\ln(1/\epsilon)$.

Although f may only make sense physically when ϵ is real and positive, we now extend our mathematical consideration of f onto the complex ϵ-plane. The radius of convergence is then the distance from the
origin to the *nearest singularity* of $f(\epsilon)$ in the complex ϵ-plane. Thus
the convergence of the expansion can be made poor by an unphysical
singularity.

If the nearest singularity is on the positive ϵ-axis, i.e. a physically real
singularity, then the signs of the coefficients c_n eventually become the

same.

$$\text{E.g.} \qquad \frac{1}{1-\epsilon} \quad \sim \quad 1 + \epsilon + \epsilon^2 + \epsilon^3$$

If the nearest singularity is on the negative ϵ-axis, then the signs eventually alternate.

$$\text{E.g.} \qquad \frac{1}{1+\epsilon} \quad \sim \quad 1 - \epsilon + \epsilon^2 - \epsilon^3$$

The *pattern of signs* is usually established quickly: it will take many terms only if the amplitude of the nearest singularity is relatively weak. If there are several singularities at the same distance, then the most singular one dominates. If there are several singularities at the same distance of equal singularity, then the one with the largest amplitude dominates. When there are several singularities of equal singularity and equal amplitude, as must happen for a real f with complex singularities necessarily occurring in complex conjugate pairs, then a complicated pattern of singularity can emerge.

$$\text{E.g.} \qquad \frac{1+\epsilon}{1+\epsilon^2} \quad \sim \quad 1 + \epsilon - \epsilon^2 - \epsilon^3 + \epsilon^4 + \epsilon^5 - \epsilon^6 - \epsilon^8$$

This example has a pattern of signs $++--$. If there are singularities in the directions $\pm\beta$ with $\beta = 2\pi M/N$, then the pattern of the signs is a cycle N long, with $2M$ changes of sign per cycle.

From elementary complex variable theory, the *radius of convergence* ϵ_0 can be calculated from

$$\epsilon_0 \quad = \quad \lim_{n\to\infty} \frac{c_{n-1}}{c_n}$$

More information, however, can be extracted from the *Domb–Sykes* plot. If there is just one nearest singularity in f at $\epsilon = \epsilon_0$ and it has an index α, i.e. f has a dominant factor

$$\begin{cases} (\epsilon_0 - \epsilon)^\alpha & \text{for } \alpha \neq 0, 1, 2, \ldots \\ (\epsilon_0 - \epsilon)^\alpha \ln(\epsilon_0 - \epsilon) & \text{for } \alpha = 0, 1, 2, \ldots \end{cases}$$

then when this factor dominates, the coefficients c_n behave like

$$\frac{c_n}{c_{n-1}} \quad \sim \quad \frac{1}{\epsilon_0}\left(1 - \frac{1+\alpha}{n}\right)$$

Thus if one plots the ratio c_n/c_{n-1} against $1/n$, the intercept gives the radius of convergence ϵ_0 and the slope gives the index of the singular-

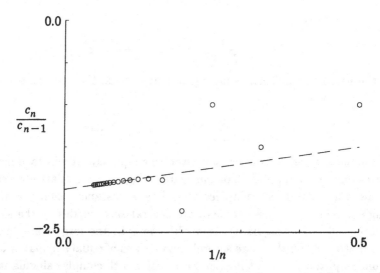

Fig. 8.1   The Domb–Sykes plot for $f(\epsilon) = \epsilon(1 + \epsilon^3)(1 + 2\epsilon)^{-1/2}$ with intercept -2 and slope 1 corresponding to $\epsilon_0 = -\frac{1}{2}, \alpha = -\frac{1}{2}$.

ity α. Figure 8.1 gives an example for

$$f(\epsilon) = \epsilon(1 + \epsilon^3)(1 + 2\epsilon)^{-1/2}$$
$$\sim \epsilon - \epsilon^2 + \tfrac{3}{2}\epsilon^3 - \tfrac{3}{2}\epsilon^4 + \tfrac{27}{8}\epsilon^5 - \tfrac{51}{8}\epsilon^6 + \tfrac{191}{16}\epsilon^7 - \tfrac{359}{16}\epsilon^8$$

By $n = 7$ the coefficients become dominated by the nearest singularity.

If one knows the value of either ϵ_0 or α from the physics, then this information can be used to help the extrapolation. Some badly bent curves can be straightened by an offset in n, i.e. plotting against $1/(n - \Delta)$. If there are several singularities at the same distance which cause the coefficients to oscillate, then one can try plotting $(c_n/c_{n-2})^{1/2}$.

If the radius of convergence is found to be infinity, then a behaviour $c_n/c_{n-1} \sim k/n$ corresponds to a factor like $\exp(k\epsilon)$, while a behaviour like $c_n/c_{n-1} \sim k/n^{1/p}$ corresponds to an integral function of order p like $\exp(\epsilon^p)$.

If the radius of convergence is found to be zero, then f has an essential singularity. This occurs with asymptotic expansions which diverge. If the coefficients behave like $c_{n-1}/c_n \sim 1/kn$, then further coefficients will be given by $c_n \sim$ constant $k^n \, n!$ as $n \to \infty$.

Exercise 8.1.   The complete elliptic integral of the first kind is

$$K(m) = \int_0^{\pi/2} \frac{d\theta}{\sqrt{1 - m^2 \cos^2 \theta}}$$

Find the general term in an expansion for small m. Use the Domb–Sykes plot to find the singularity nearest to the origin and its type.

Exercise 8.2. Find the general term in an expansion for small k for the integral

$$\int_0^1 \frac{dx}{(1 + kx^2)^{1/3}}$$

Use the Domb–Sykes plot to find the singularity nearest to the origin and find its type.

8.2 Improved series

There is no one best method. The following seven techniques are appropriate to different situations.

Reversion

A singularity on the positive axis usually means that $f(\epsilon)$ is multivalued, and that there is a maximum attainable ϵ. Often the inverted function

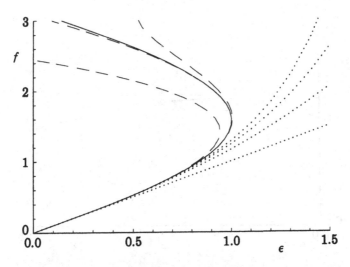

Fig. 8.2   Inversion. The continuous curve is $f(\epsilon) = \sin^{-1}\epsilon$. The dotted curves are increasingly higher order expansions of f in terms of ϵ. The dashed curves are increasingly higher order expansions of the inversion, $\epsilon(f)$.

$\epsilon(f)$ is single valued. Thus recasting the result as a series in f for ϵ can give an expression which continues onto the upper branch.

Figure 8.2 gives an example for

$$f(\epsilon) \;=\; \sin^{-1}\epsilon$$
$$\sim\; \epsilon + \tfrac{1}{6}\epsilon^3 + \tfrac{3}{40}\epsilon^5 + \tfrac{5}{112}\epsilon^7$$

with inversion

$$\epsilon \;\sim\; f - \tfrac{1}{6}f^3 + \tfrac{1}{120}f^5 - \tfrac{1}{5040}f^7$$

which gives a good approximation on to the next branch.

Taking a root

If f has a mild singularity such as a branch cut, i.e.

$$f \;\sim\; A(\epsilon_0 - \epsilon)^\alpha \quad \text{as } \epsilon \to \epsilon_0 \quad \text{with } \alpha > 0$$

then the appropriate root of f, $f^{1/\alpha}$, does not have this singularity at ϵ_0.

Figure 8.3 gives an example for

$$f(\epsilon) \;=\; e^{-\epsilon/2}\sqrt{1+2\epsilon}$$
$$\sim\; 1 + \tfrac{1}{2}\epsilon - \tfrac{7}{8}\epsilon^2 + \tfrac{41}{48}\epsilon^3 - \tfrac{367}{384}\epsilon^4 + \tfrac{4849}{3840}\epsilon^5$$

which has a radius of convergence of $\tfrac{1}{2}$, whereas

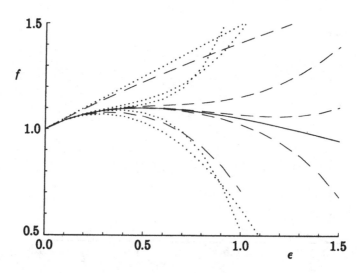

Fig. 8.3   Taking a root. The continuous curve gives $f(\epsilon) = e^{-\epsilon/2}\sqrt{1+2\epsilon}$. The dotted curves are increasingly higher order expansions of f in terms of ϵ. The dashed curves are increasingly higher order expansions of f^2 in terms of ϵ.

$$f^2 \quad \sim \quad 1 + \epsilon - \tfrac{3}{2}\epsilon^2 + \tfrac{5}{6}\epsilon^3 - \tfrac{7}{24}\epsilon^4 + \tfrac{3}{40}\epsilon^5$$

has an infinite radius of convergence.

Multiplicative extraction

From a Domb–Sykes plot one can know that $f(\epsilon)$ has a nearest singularity at $\epsilon = \epsilon_0$ with an index α. If this singularity is factored out multiplicatively, by setting

$$f(\epsilon) \;=\; (\epsilon_0 - \epsilon)^\alpha f_M(\epsilon)$$

then the new function f_M should be regular at $\epsilon = \epsilon_0$, and ought to have singularities further away from the origin, i.e. ought to converge better.

Additive extraction

A singular factor can be extracted additively instead of multiplicatively if the amplitude A is also known:

$$f(\epsilon) \;=\; A(\epsilon_0 - \epsilon)^\alpha + f_A(\epsilon)$$

If the singular factor is known to be additive and if $\alpha > 0$, then a multiplicative extraction would fail to produce better behaviour in f_M, e.g.

$$1 + (1 - \epsilon)^{1/2} = (1 - \epsilon)^{1/2} f_M \quad \text{with} \quad f_M = 1 + (1 - \epsilon)^{-1/2}$$

Euler transformation

A non-physical singular point $\epsilon_0 > 0$ can be transformed to infinity by introducing a new small parameter

$$\tilde{\epsilon} \;=\; \frac{\epsilon}{1 - \epsilon/\epsilon_0}$$

The series expansion of f recast in terms of $\tilde{\epsilon}$, $f \sim \sum d_n \tilde{\epsilon}^n$, has the non-physical singularity pushed out to $\tilde{\epsilon} = \infty$. Hence the new series in $\tilde{\epsilon}$ should converge better on the physical real positive $\tilde{\epsilon}$-axis.

Figure 8.4 gives an example for

$$f(\epsilon) \;=\; e^{-\epsilon/2}\sqrt{1 + 2\epsilon}$$

This has a non-physical singularity at $\epsilon = -\tfrac{1}{2}$, which can be mapped to ∞ with the Euler transform $\tilde{\epsilon} = \epsilon/(1 + 2\epsilon)$. The expansion of f in

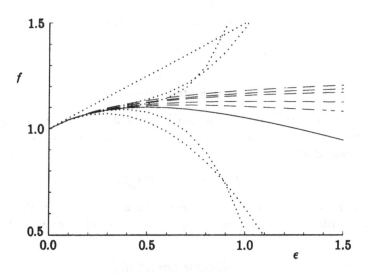

Fig. 8.4 Euler transformation. The function $f(\epsilon) = e^{-\epsilon/2}\sqrt{1+2\epsilon}$ is given by the continuous curve. The dotted curves are for increasingly higher order expansions of f in terms of ϵ. The dashed curves are for increasingly higher order expansions of f in terms of $\bar{\epsilon}$.

terms of $\bar{\epsilon}$ is

$$f \;\sim\; 1 + \tfrac{1}{2}\bar{\epsilon} + \tfrac{1}{8}\bar{\epsilon}^2 - \tfrac{31}{48}\bar{\epsilon}^3 - \tfrac{895}{384}\bar{\epsilon}^4 - \tfrac{22591}{3840}\bar{\epsilon}^5$$

While the transformed expansion does not have the wild behaviour of the original expansion at $\epsilon = \tfrac{1}{2}$, it does not provide a particularly good approximation beyond the old radius of convergence.

Shanks transform

This transform assumes that the partial sums of n terms, $S_n = \sum^n c_k \epsilon^k$, are in a geometric progression

$$S_n \;=\; A + BC^n$$

The answer A can then be extracted from just three partial sums by a nonlinear extrapolation

$$A \;=\; \frac{S_{n+1} S_{n-1} - S_n^2}{S_{n+1} - 2S_n + S_{n-1}} \;=\; S_n - \frac{(S_{n+1} - S_n)(S_n - S_{n-1})}{(S_{n+1} - S_n) - (S_n - S_{n-1})}$$

(The second form is more stable for computations.) This extrapolation often works better than it should do. Note that it will work in diverging series, when $|C| > 1$. It can be used repeatedly, i.e. the A's produced from $n = 1, 2, 3, \ldots$ can themselves be considered a series of partial sums to which the Shanks transform can be applied resulting in new improved A's. Such repeated applications of the transform effectively remove further geometric terms in $S_n = A + BC^n + DE^n + FG^n + \cdots$, with $|C| > |E| > |G|$.

As an example consider the slowly converging series

$$\ln 2 \;=\; 1 - \tfrac{1}{2} + \tfrac{1}{3} - \tfrac{1}{4} + \tfrac{1}{5} - \tfrac{1}{6} + \tfrac{1}{7} - \tfrac{1}{8} + \tfrac{1}{9} + \cdots$$

The table below gives the results from repeated applications of the Shanks transform

Partial sums	1-Shanks	2-Shanks	3-Shanks	4-Shanks	Exact
1.000					
0.500					
0.833	0.7000				
0.583	0.6905				
0.783	0.6944	0.693277			
0.617	0.6924	0.693106			
0.760	0.6936	0.693163	0.6931489		
0.635	0.6929	0.693140	0.6931467		
0.746	0.6933	0.693151	0.6931474	0.693147196	0.69314718056

Using the first, less stable, form of the Shanks transform and working with 11 figure accuracy will produce 0.693147404 rather than 0.693147196 in the single result for the fourth application of the Shanks transform.

Exercise 8.3 Repeatedly apply the Shanks transform to just the 6 terms

$$1 - \frac{1}{3} + \frac{1}{5} - \frac{1}{7} + \frac{1}{9} - \frac{1}{11}$$

to obtain an estimate for $\pi/4$ with an error of 2.10^{-4}.

Padé approximants

This is a very popular improvement. The expansion is re-expressed as a rational polynomial

$$f(\epsilon) \;\sim\; \sum_0^{M+N} c_n \epsilon^n \;\sim\; \frac{\sum_0^M a_n \epsilon^n}{\sum_0^N b_n \epsilon^n}$$

This is called the $[M/N]$ approximant. In general the diagonal case $M = N$ is best, unless one has some special knowledge.

There are various ways of obtaining the coefficients a_n and b_n. The above expression can be multiplied by the denominator $\sum b_n \epsilon^n$ and the result truncated after ϵ^{M+N} to yield

$$
\begin{aligned}
b_0 c_0 &+ (b_0 c_1 + b_1 c_0)\epsilon \\
&+ \cdots + (b_0 c_N + \cdots + b_N c_0)\epsilon^N + (b_0 c_{N+1} + \cdots + b_N c_1)\epsilon^{N+1} \\
&+ \cdots + (b_0 c_{M+N} + \cdots + b_N c_M)\epsilon^{N+M} \\
= a_0 &+ a_1 \epsilon \\
&+ \cdots + a_M \epsilon^M + 0\epsilon^{M+1} \\
&+ \cdots + 0\epsilon^{M+N}
\end{aligned}
$$

Comparing the coefficients of ϵ^n for $n = M + 1$ to $M + N$ gives a square matrix to invert for the $b_1, \ldots, b_N$ if one takes $b_0 = 1$. The a_n are then read off from ϵ^n for $n = 0$ to M. One disadvantage of this method is that all the coefficients a_n and b_n change when the degrees M and N are changed. An alternative method uses a continued fraction representation

$$
f(\epsilon) \quad = \quad \cfrac{d_0}{1 + \cfrac{d_1 \epsilon}{1 + \cfrac{d_2 \epsilon}{1 + \cfrac{d_3 \epsilon}{1 + \cdots}}}}
$$

In this representation the values of the coefficients d_n do not depend on the truncation point.

The Padé approximants have many interesting properties. So long as M and N are sufficiently large, a pole in f is represented by a pole in $[M/N]$ at nearly the correct position. An essential singularity in f is represented by an increasing cluster of poles nearby. A branch cut in f is represented by a line of poles and zeros radiating out from the branch point along a line through the origin. If there are non-polar singularities, it can be better to form a Padé approximant to the logarithmic derivative of the function.

As an example consider

$$
\ln(1 + \epsilon) \quad = \quad \epsilon - \tfrac{1}{2}\epsilon^2 + \tfrac{1}{3}\epsilon^3 - \tfrac{1}{4}\epsilon^4 + \tfrac{1}{5}\epsilon^5 + \cdots
$$

This has a continued fraction representation

$$
\epsilon/(1 + \tfrac{1}{2}\epsilon/(1 + \tfrac{1}{6}\epsilon/(1 + \tfrac{1}{3}\epsilon/(1 + \tfrac{1}{5}\epsilon/(1 + \tfrac{3}{10}\epsilon/(1 + \tfrac{3}{14}\epsilon/(1+
$$
$$
\tfrac{2}{7}\epsilon/(1 + \tfrac{2}{9}\epsilon/(1 + \tfrac{5}{18}\epsilon/(1 + \tfrac{5}{22}\epsilon/(1 + \cdots))))))))))
$$

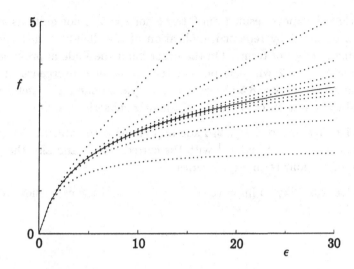

Fig. 8.5   The Padé approximants for $\ln(1 + \epsilon)$ from the continued fraction representation.

The above truncation has poles and zeros scattered along the negative real ϵ-axis up to $\epsilon = -1$

poles -1.041 -1.247 -1.779 -3.284 -10.15

zeros -1.079 -1.409 -2.365 -6.253 -123.3

Successive truncations of the continued fraction evaluated at $\epsilon = 1$ and at $\epsilon = 2$ yield

	$\epsilon = 1$	$\epsilon = 2$
	1.000	2.000
	0.667	1.000
	0.700	1.143
	0.6923	1.091
	0.69333	1.1014
	0.693122	1.0980
	0.693152	1.0988
	0.6931464	1.09857
	0.6931473	1.098626
	0.693147158	1.098609
	0.6931471850	1.0986132
	...	...
exact	0.69314718056	1.0986122887

Note that the above result from 9 terms for $\epsilon = 1$ is not quite as good as that obtained by repeated application of the Shanks transform on the same number of terms. On the other hand the Padé approximants converge at $\epsilon = 2$ which is beyond the radius of convergence of the original series. Figure 8.5 shows that the approximants are useful well beyond the radius of convergence of the original series.

Exercise 8.4   Find the [2/2] Padé approximant to $\exp(x)$. Compare the predictions for e^1 and e^{-1} with the correct values and also the prediction of the four term Taylor series.

Exercise 8.5   Try to improve the convergence of the expansion in exercise 8.2.

Bibliography

Baker, G.A. (1975). *Essentials of Padé Approximants* Academic Press.

Bender, C.M. & Orszag, S.A. (1978). *Advanced mathematical methods for scientists and engineers* McGraw-Hill.

Brillouin, L. (1926). Remarques sur la méchanique ondulatoire. *J. Phys. Radium* **7**, 353–368.

de Bruijn, N.G. (1958). *Asymptotic methods in analysis* North-Holland.

Cole, J.D. (1968). *Perturbation methods in applied mathematics* Ginn-Blaisdell.

Copson, E.T. (1965). *Asymptotic expansions* Cambridge University Press.

Domb, C. & Sykes, M.F. (1957). On the susceptibility of a ferromagnetic above the Curie point. *Proc. R. Soc. Lond.* A **240**, 214–228.

Eckhaus, W. (1979). *Asymptotic analysis of singular perturbations* North-Holland.

Erdélyi, A. (1956). *Asymptotic expansions* Dover.

Fraenkel, L.E. (1969). On the method of matched asymptotic expansions, Parts I, II and III. *Proc. Camb. Phil. Soc.* **65**, 209–284.

Gans, R. (1915). Propagation of light through an inhomogeneous medium. *Ann. Phys.* **47**, 709–736.

Green, G. (1837). On the motion of waves in a variable canal of small depth and width. *Trans. Camb. Phil. Soc.* **6**, 457–462.

Horn, J. (1899). Untersuchung der Integrale einer linearen Differentialgleichung in der Umgebung einer Unbestimmtheitsstelle vermittelst successiver Annäherungen. *Arch. Math. Physik. Leipzig* **4**, 213–230.

Jeffreys, H. (1924). On certain approximate solutions of linear differential equations of the second order. *Proc. Lond. Math. Soc.* **23**, 428–436.

Kaplun, S. (1957). Low Reynolds number flow past a circular cylinder *J. Math. Mech.* **6**, 595–603.

Kaplun, S. (1967). *Fluid mechanics and singular perturbations* ed. P.A. Lagerstrom, L.N. Howard & C.S. Liu. Academic Press.

Kevorkian, J. & Cole, J.D. (1981). *Perturbation methods in applied mathematics* Springer–Verlag.

Kramers, H.A. (1926). Wellenmechanik und halbzahlige Quantisierung. *Z. Phys.* **38**, 828–840.

Lagerstrom, P.A. & Casten, R.G. (1972). Basic concepts underlying singular perturbation techniques. *SIAM Rev.* **14**, 63–120.

Langer, R.E. (1931). On the asymptotic solution of ordinary differential equations, with an application to the Bessel functions of large order. *Trans. Am. Math. Soc.* **33**, 23–64.

Langer, R.E. (1935). On the asymptotic solution of ordinary differential equations, with reference to the Stokes' phenomenon about a singular point. *Trans. Am. Math. Soc.* **37**, 397–416.

Lighthill, M.J. (1949). A technique for rendering approximate solutions to physical problems uniformly valid. *Phil. Mag.* **40**, 1179–1201.

Lin, C.C. (1954). On a perturbation theory based on the method of characteristics. *J. Maths & Phys.* **33**, 117–134.

Lin, C.C. & Segel, L.A. (1974). *Mathematics applied to deterministic problems in the natural sciences* Macmillan.

Liouville, J. (1837). Sur le développement des fonctions ou parties de fonctions en séries.... *J. Math. Pures Appl.* **2**, 16–35.

Meyer, R.E. (1975). Gradual reflection of short waves. *SIAM J. Appl. Math.* **29**, 481–492.

Nayfeh, A.H. (1973). *Perturbation Methods* John Wiley & Sons.

Olver, F.W.J. (1961). Error bounds for the Liouville–Green (or WKB) approximation *Proc. Camb. Phil. Soc.* **57**, 790–810.

Olver, F.W.J. (1974). *Introduction to asymptotics and special functions* Academic Press.

O'Malley, R.E. (1974). *Introduction to singular perturbations* Academic Press.

Padé, H. (1892). *Sur la représentation approachée d'une fonction pour des fractions rationelles* Thesis École Normal Sup.

van der Pol, B. (1926). On 'relaxation–oscillations'. *Phil. Mag.* **2**, 978–992.

Proudman, I. & Pearson, J.R.A. (1957). Expansions at small Reynolds numbers for the flow past a sphere and a cylinder. *J. Fluid Mech.* **2**, 237–262.

Rayleigh, Lord (1912). On the propagation of waves through a stratified medium, with special reference to the question of reflection. *Proc. R. Soc. Lond.* A **86**, 207–226

Shanks, D. (1955). Nonlinear transforms of divergent and slowly convergent sequences *J. Maths & Phys.* **34**, 1–42.

Van Dyke, M. (1964). *Perturbation methods in fluid mechanics* Academic Press. Annotated version (1975) Parabolic Press.

Van Dyke, M. (1974). Analysis and improvement of perturbation series. *Q. J. Mech. Appl. Math.* **27**, 423–450.

Van Dyke, M. (1975). Computer extension of perturbation series in fluid mechanics. *SIAM J. Appl. Maths* **28**, 720–734.

Wasow, W. (1965). *Asymptotic expansions for ordinary differential equations* Interscience.

Wentzel, G. (1926). Eine Verallgemeinerung der Quantenbedingungen für die Zwecke der Wellenmechanik. *Z. Phys.* **38**, 518–529.

Index